Physical Biology of Proteins and Peptides

Luis Olivares-Quiroz • Orlando Guzmán-López
Héctor Eduardo Jardón-Valadez
Editors

Physical Biology of Proteins and Peptides

Theory, Experiment, and Simulation

Springer

Editors
Luis Olivares-Quiroz
Autonomous University of Mexico City
Mexico City, Mexico

Héctor Eduardo Jardón-Valadez
Autonomous Metropolitan University
(Lerma), Lerma de Villada
Mexico City, México

Orlando Guzmán-López
Autonomous Metropolitan University
(Iztapalapa)
Mexico City, Mexico

ISBN 978-3-319-21686-7 ISBN 978-3-319-21687-4 (eBook)
DOI 10.1007/978-3-319-21687-4

Library of Congress Control Number: 2015947775

Springer Cham Heidelberg New York Dordrecht London

Printed on acid-free paper

Springer International Publishing AG Switzerland is part of Springer Science+Business Media (www.springer.com)

Preface

Interdisciplinary approaches emerge as promising methodologies for understanding relevant biological systems at different spatial and temporal scales. Because of the structural and dynamical complexity found in these systems, traditional conceptions about the principles that govern the behavior of living organisms have been challenged. Hence, in order to tackle the great scientific puzzles faced by modern science at the interface of Physics, Molecular Biology, and Chemistry, a full and new interdisciplinary research has to be launched.

It is within this context that the first conference *Physical Biology of Proteins and Peptides: Theory, Experiment and Simulation* took place in Mexico City in early 2015. Experts from the USA, United Kingdom, European Union, Germany, Russia, and Latin America gathered in the heart of Mexico City to discuss novel and exciting approaches to discuss the new trends of the macroscopic behavior of biological macromolecules. The meeting gathered approximately 20 speakers and a considerable large audience of undergraduate and graduate students coming from the fields of molecular biology, organic chemistry, physics and computational sciences. The conference fulfilled all our goals and expectations greatly. Attendees could capture some of the current trends in the Physical Biology field, in particular strategies combining theoretical and experimental studies, and validation of models set up in computer simulations.

These proceedings represent the collective effort of Mexican and international scientists and institutions for developing Physical Biological sciences with an interdisciplinary approach. The overwhelming positive response of researchers and students attending the meeting testifies to the number of labs working in the field in Mexico. We hope that this collective initiative serves as a starting point to trigger larger collaborations among research groups based in Mexico and abroad, from basic principles to clinical and biotechnological applications, such as the examples described in the following chapters. Even if the topics are diverse, authors were able to address microscopic determinants, some of them at the atomistic level, that are linked to a functional role of proteins and peptides, including its dynamic environment.

First, Del Pozo-Yauner and collaborators describe protein aggregation in human amyloidosis; specifically, they discuss the structural determinants of fibril formation. Next, in a computational experiment using all-atom molecular dynamics simulations, Barz and Strodel analyze the kinetics of aggregations for the amyloid β-peptide, including alloforms with 40 and 42 amino acids. In the third chapter, Finkelstein and collaborators model fibril formation in two main regimes, namely, primary nucleation with linear growth and exponential growth with branching, fragmentation, and pronounced lag times. Then, Taylor et al. analyze the role of the Hsp90 chaperone in the cholera toxin infection mechanism, suggesting that it promotes the toxin's functional refolding in the cytosol. Afterwards, Rivillas-Acevedo, Fernández-Silva, and Amero review the conditions for the onset of cataract formation, and suggest strategies for preventing them. In chapter "Protein Motions, Dynamic Effects and Thermal Stability in Dihydrofolate Reductase from the Hyperthermophile Thermotoga maritima," using dihydrofolate reductase as model system, Allemann, Loveridge, and Luk analyze the link between the enzyme conformational dynamics and the reaction kinetics. After this, Jardón-Valadez, Bond, and Ulmschneider perform molecular simulations to characterize the folding of glycophorin A and the acetylcholine M2 domain. Subsequently, Townsend and collaborators provide evidence of allosteric regulation of enzymes by specific control sites. In chapter "Dynamics of Membrane Proteins and Lipid Bilayers," Jardón-Valadez provides an overview of the structure and dynamics of lipid bilayers and embedded proteins, by comparing results from computer simulation and neutron diffraction experiments. Finally, Hernández-Lemus, Baca-López, and Tovar analyze the role of a transcription factor as master regulator.

A highlighted feature of this conference was the Round Table organized on the second day of the meeting. Reflections and opinions poured during this activity captured the excitement and novelty of this field. Physical Biology understood as the set of approaches and tools imported from Physics and Biology was suggested, analyzed, and debated, as well as the recognition that complexity in living organisms requires an open discussion of the validity of fundamental principles beyond particular disciplinary views.

In launching new enterprises, words may not be relevant just because of their definition, but due to the goals and purposes they embody. Therefore, we hope that this volume conveys that Physical Biology and Biological Physics address the inherent complexity of living organisms, as well as the need of building bridges among disciplines to attain this goal. We are confident that this book shall provide the reader with a set of relevant and cutting-edge research topics at the interface of Physics, Molecular Biology, Chemistry, and Computational Sciences.

Finally, we would like to acknowledge all financial and administrative support received from our Institutions. We are grateful to Universidad Autonoma de la Ciudad de Mexico (UACM), Universidad Autonoma Metropolitana-Lerma (UAM-L), Universidad Autonoma Metropolitana Iztapalapa (UAM-I), and CINVESTAV-Zacatenco for all their support. In addition, this effort could not have been possible without the enthusiastic contributions from our colleagues.

Mexico City, Mexico — Luis Olivares-Quiroz
Mexico City, Mexico — Orlando Guzmán-López
Lerma de Villada, Edo de México — Héctor Eduardo Jardón-Valadez

Contents

Contributors

Rudolf K. Allemann School of Chemistry, Cardiff University, Cardiff, UK

Carlos Amero Centro de Investigaciones Químicas, Universidad Autónoma del Estado de Morelos, Morelos, Mexico

Karol Baca-López Tianguistenco Higher Studies Institute of Technology, Santiago Tianguistenco, Mexico

Bogdan Barz Forschungszentrum Jülich, Institute of Complex Systems: Structural Biochemistry (ICS-6), Jülich, Germany

Baltazar Becerril Instituto de Biotecnología UNAM, Mexico

Peter J. Bond Bioinformatics Institute, Singapore, Singapore

Christopher B.T. Britt Burnett School of Biomedical Sciences, College of Medicine, University of Central Florida, Orlando, FL, USA

Martin J. Cann Biophysical Sciences Institute, Durham University, Durham, UK

School of Biological and Biomedical Sciences, Durham University, Durham, UK

Julio Isael Pérez Carrión Laboratorio de Estructura de Proteínas, Instituto Nacional de Medicina Genómica, México D.F., Mexico

Nikita V. Dovidchenko Institute of Protein Research, Russian Academy of Sciences, Moscow Region, Russia

Arline Fernández-Silva Centro de Investigaciones Químicas, Universidad Autónoma del Estado de Morelos, Morelos, Mexico

Alexey V. Finkelstein Institute of Protein Research, Russian Academy of Sciences, Moscow Region, Russian Federation

Faculty of Biology, Moscow State Lomonosov University, Moscow, Russian Federation

Jennifer Fundora Burnett School of Biomedical Sciences, College of Medicine, University of Central Florida, Orlando, USA

Oxana V. Galzitskaya Institute of Protein Research, Russian Academy of Sciences, Moscow Region, Russian Federation

Enrique Hernández-Lemus Center for Complexity Sciences, National Autonomous University of México, Ciudad Universitaria, México

National Institute of Genomic Medicine, México

Eduardo V. Jardon School of Chemistry, Cardiff University, Cardiff, UK

Héctor Eduardo Jardón-Valadez Departamento de Recursos de la Tierra, Autonomous Metropolitan University (Lerma), Lerma de Villada, Edo de México, México

E. Joel Loveridge School of Chemistry, Cardiff University, Cardiff, UK

Louis Y.P. Luk School of Chemistry, Cardiff University, Cardiff, UK

Tom C.B. McLeish Department of Physics, Durham University, Durham, UK

Biophysical Sciences Institute, Durham University, Durham, UK

Department of Chemistry, Durham University, Durham, UK

Adrián Ochoa-Leyva Laboratorio de Estructura de Proteínas, Instituto Nacional de Medicina Genómica, México D.F., Mexico

Ehmke Pohl School of Biological and Biomedical Sciences, Durham University, Durham, UK

Biophysical Sciences Institute, Durham University, Durham, UK

Department of Chemistry, Durham University, Durham, UK

Luis Del Pozo-Yauner Laboratorio de Estructura de Proteínas, Instituto Nacional de Medicina Genómica, México D.F., Mexico

Lina Rivillas-Acevedo Centro de Investigaciones Químicas, Universidad Autónoma del Estado de Morelos, Morelos, Mexico

Thomas L. Rodgers School of Chemical Engineering and Analytical Sciences, University of Manchester, Manchester, UK

Biophysical Sciences Institute, Durham University, Durham, UK

Department of Chemistry, Durham University, Durham, UK

Sandra Leticia Rodríguez-Ambriz Centro de Desarrollo de Productos Bióticos, IPN, Mexico

Rosana Sánchez-López Instituto de Biotecnología UNAM, Mexico

Olga M. Selivanova Institute of Protein Research, Russian Academy of Sciences, Moscow Region, Russian Federation

Birgit Strodel Forschungszentrum Jülich, Institute of Complex Systems: Structural Biochemistry (ICS-6), Jülich, Germany

Alexey K. Surin Institute of Protein Research, Russian Academy of Sciences, Pushchino, Moscow Region, Russian Federation

State Research Center for Applied Microbiology and Biotechnology, Russian Federation

Maria Yu. Suvorina Institute of Protein Research, Russian Academy of Sciences, Moscow Region, Russian Federation

Michael Taylor Burnett School of Biomedical Sciences, College of Medicine, University of Central Florida, Orlando, FL, USA

Ken Teter Burnett School of Biomedical Sciences, College of Medicine, University of Central Florida, Orlando, FL, USA

Hugo Tovar National Institute of Genomic Medicine, México

Philip D. Townsend Biophysical Sciences Institute, Durham University, Durham, UK

School of Biological and Biomedical Sciences, Durham University, Durham, UK

Martin B. Ulmschneider Department of Materials Science and Engineering, Johns Hopkins University, Baltimore, MD, USA

Institute of BioNanoTechnology, Johns Hopkins University, Baltimore, MD, USA

Daniel Alejandro Fernández Velasco Facultad de Medicina UNAM, Mexico

Mark R. Wilson Biophysical Sciences Institute, Durham University, Durham, UK

Department of Chemistry, Durham University, Durham, UK

Guadalupe Zavala-Padilla Instituto de Biotecnología UNAM, Mexico

The Structural Determinants of the Immunoglobulin Light Chain Amyloid Aggregation

Luis Del Pozo-Yauner, Baltazar Becerril, Adrián Ochoa-Leyva, Sandra Leticia Rodríguez-Ambriz, Julio Isael Pérez Carrión, Guadalupe Zavala-Padilla, Rosana Sánchez-López, and Daniel Alejandro Fernández Velasco

Abstract The extracellular deposition of a monoclonal immunoglobulin light chain (LC) as insoluble fibrillar aggregates is the cause of the primary systemic amyloidosis, also known as light chain-derived (AL) amyloidosis. In this chapter the structural factors determining the potential of the LC to aggregate into amyloid fibrils are analyzed. In the first two sections, the relationship between protein aggregation and disease, as well as the causes and characteristics of the human amyloidoses are discussed. The role of protein misfolding in the fibril assembly pathway and the structural bases of the amyloid fibrils stability are also addressed. Then, the chapter focuses on the molecular causes of the amyloid aggregation of LC. The contribution of the somatic mutations and the identity of the variable gene segment encoding the LC to the propensity to form amyloid fibrils are discussed. The chapter summarizes the experimental evidence supporting the role of partially folded intermediates in the mechanism of AL fibril assembly, as well as the modulatory effect played by the variable region gene sequence on the aggregation behavior of the LC. Finally, the results of a recent study aimed to identify fibrillogenic regions in the λ6 recombinant variable domain protein 6aJL2 are presented.

Keywords Misfolding disease • Amyloidosis • Amyloid fibril • Protein misfolding • Non-native intermediate • Protein aggregation • Immunoglobulin light chain • Lambda 6 subgroup • Fibrillogenesis • Profibrillogenic sequence

---

L. Del Pozo-Yauner (✉) • A. Ochoa-Leyva • J.I.P. Carrión
Laboratorio de Estructura de Proteínas, Instituto Nacional de Medicina Genómica, Periférico Sur No. 4809, Col. Arenal Tepepan, Tlalpan, México D.F. 14616, Mexico
e-mail: ldelpozo@inmegen.gob.mx

B. Becerril • G. Zavala-Padilla • R. Sánchez-López
Instituto de Biotecnología UNAM, Mexico

S.L. Rodríguez-Ambriz
Centro de Desarrollo de Productos Bióticos, IPN, Mexico

D.A.F. Velasco
Facultad de Medicina UNAM, Mexico

L. Olivares-Quiroz et al. (eds.), *Physical Biology of Proteins and Peptides*, DOI 10.1007/978-3-319-21687-4_1

Protein Aggregation as a Cause of Disease

Proteins determine the phenotype in all biological systems; from unicellular bacteria and yeast, to the more complex multicellular organisms, like humans, as they are responsible for translating the information encoded in the genome into specific molecular actions [1, 2]. The functional properties of a protein correspond to its native state, which is generally described as a relatively stable and compact 3D conformation reached by the molecule through the folding process [3]. Protein folding runs almost parallel to ribosomal synthesis and involves several specialized proteins called molecular chaperones, which work coordinately to ensure that the nascent polypeptide chain reaches its native conformation without risk to the host cell [4]. However, given the challenging conditions of the intracellular environment; characterized by a high concentration of molecules of very diverse structural and physicochemical properties; the probability that at least a fraction of the molecules of a protein get trapped in an intermediate state of the folding pathway is relatively high [5]. The failure of efficient folding, or the loss of the native state after proper folding, leads often to protein aggregation. Once formed, these aggregates can actively interact with other cellular components, causing disruption of normal organelle functions. Thus, protein aggregation implies a fatal risk for the cell, either if aggregation occurs intracellularly, or in the extracellular milieu [6, 7]. The cell relies largely on the molecular chaperones of the Protein Quality Control (PQC) system and the proteasome complex for the safe elimination of the misfolded and aggregated proteins that accumulate intracellularly, ensuring the conservation of the proteostasis [8]. There is increasing evidence about the protective function of a group of extracellular proteins with chaperone activity, which are probably part of a protective mechanism homologous to the intracellular PQC system [9]. Nevertheless, under certain cellular conditions, the rate of protein synthesis can increase in such a way that the capacity of the systems responsible for proper folding of newly synthesized proteins is exceeded. The consequence is usually the accumulation of misfolded proteins and their aggregates. A similar result may also arise because of the progressive decline of the cellular functions that usually accompany aging [6]. Today, it is well established that the pathogenesis of a clinically heterogeneous group of human diseases, including certain types of cancers [10, 11], metabolic imbalances [12, 13], and neurodegenerative diseases [12], is directly or indirectly related to the protein aggregation. These pathologies have been classified under the general term “protein misfolding diseases” and presently represent a very active field of research [14]. Given the essential role played by proteins in all aspects of cell biology, it can be anticipated that, in the near future, there will be an increase in the number of illnesses in which a causal relationship with protein aggregation can be substantiated. This increase will parallel our improved ability to detect and characterize subtle events of protein misfolding and aggregation *in vivo*.

Proteins can form different types of aggregates *in vivo*, which differ by their appearance under the electron microscope, their biophysical properties, and their pathogenic potential. One of the types of pathogenic aggregates that proteins can form is the amyloid fibril, which is the main component of the deposits appearing in the course of amyloidoses [15].

Human Amyloidoses

Amyloidosis is the term that encompasses a group of clinically heterogeneous diseases, characterized by deposits in the extracellular compartment of one or more organs of a substance of complex composition termed "amyloid". This group includes the type II diabetes mellitus, Parkinson's and Alzheimer's diseases, transmissible spongiform encephalopathies—also known as prion diseases—, and primary systemic amyloidosis, among others [16]. The specific component of the amyloid deposits are fibrillar aggregates formed by a protein or peptide; whose molecular identity is distinctive of each disease. Currently, about 25 different proteins and peptides have been identified as capable of forming amyloids in humans. They are known as amyloid precursors and widely differ in primary structure and 3D conformation (Fig. 1). Despite this, they can form aggregates that share much of their morphological, spectroscopic, and tinctorial properties, reflecting the structural similarity that links all the amyloids [17, 18] (Fig. 2).

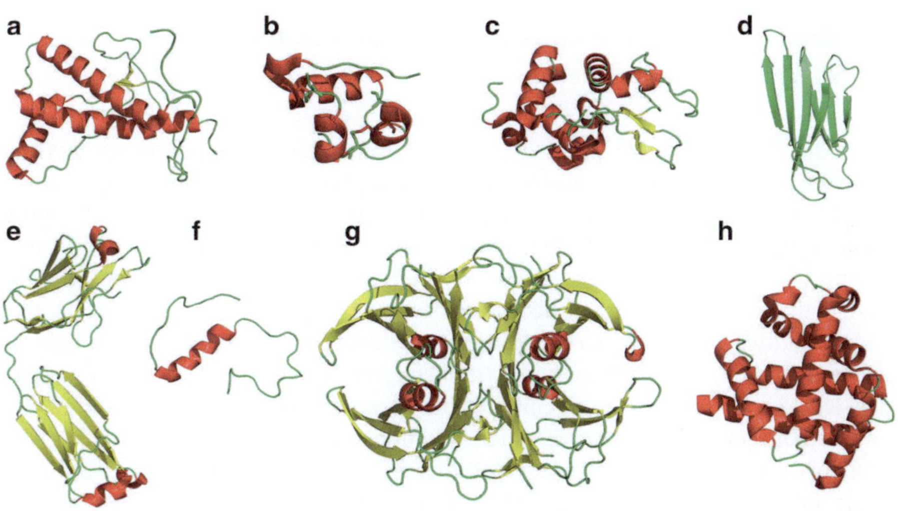

Fig. 1 Structural diversity of human amyloid precursors. Cartoon representation of the 3D structure of (**a**) the protective variant E219K of the prion protein (PDB ID 2LSB), (**b**) insulin (PDB ID 2C8Q), (**c**) lysozyme (PDB ID 1IWT), (**d**) β_2 microglulin (PDB ID 2F8O), (**e**) immunoglobulin light chain of λ type (PDB ID 3WHE), (**f**) Aβ(1-40) peptide in an aqueous environment (PDB ID 2LFM), and (**g**) the highly amyloidogenic transthyretin mutant G53S/E54D/L55S (PDB ID 1G1O). In (**h**) the crystallographic structure of the human myoglobin mutant K45R (PDB ID 3RGK) is shown. Myoglobin forms amyloid-like aggregates in in vitro conditions, but is not related to in vivo amyloid deposition. All the structures were determined by X-ray crystal diffraction, with the exception of those shown in (**a**, **f**), which were determined by solution NMR. The molecular graphics figures were prepared with PyMOL™

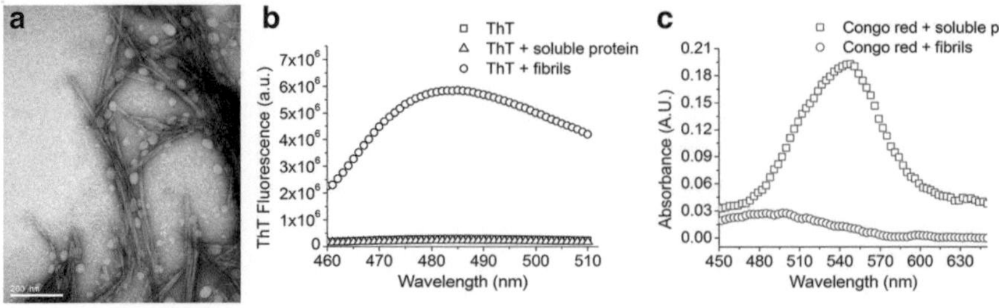

Fig. 2 Common morphological and spectroscopic properties of amyloids. (**a**) Amyloid-like fibrils formed in vitro by the mutant F2L of the recombinant light chain variable domain (rV_L) protein 6aJL2 [19]. The amyloid deposits, regardless to the chemical composition, are characteristically composed by unbranched fibrils of variable length, but with a diameter in the range of 7–10 nm. (**b**) Fluorescence emission, and (**c**) visible absorption spectra of, respectively, thioflavin T (ThT) and Congo red in the presence of the amyloid-like aggregates shown in *panel* A. The fluorescence emission spectrum of the ThT bound to amyloid fibrils is characterized by a λ_{max} of around 482 nm, when it is excited at 450 nm. The visible absorption spectra of the dye Congo red shows a shift of the absorption maximum from 490 nm, characteristics of the spectra of the free dye or in the presence of soluble protein, to 541 nm in the presence of amyloid aggregates

The Role of Partially Folded Intermediates in the Fibril Assembly Mechanism

From the thermodynamic point of view, the spontaneous formation of amyloid fibrils by a globular protein proves that, under appropriated conditions, a polypeptide chain can adopt a tridimensional conformation that is thermodynamically more stable than the native one [20, 21]. This is a relevant issue for the theory of protein folding, given that the native state was considered for a long time as the global minimum in the energetic landscape accessible for a polypeptide chain with a productive folding pathway [22]. An increasing amount of data about the structural and physicochemical properties of the amyloid fibril supports the concept that the native state is, at least in the case of the proteins capable to self-assembling into amyloid fibrils, a metastable conformation that is kinetically trapped in a local minimum, due to the relatively large energetic barrier separating it from the thermodynamically most stable amyloid conformation [20, 21]. A fundamental issue, whose methodological approach is usually challenging, is the elucidation of the molecular mechanism by which a protein transits from the soluble native state to the insoluble fibrillar. It is well known that destabilizing mutations [23, 24], as well as stressing conditions, such as acidic pH [25, 26], high temperature [27], or the addition of chaotropic salts, such as urea and guanidine hydrochloride (GdnHCl) [28], accelerate the *in vitro* conversion of proteins into amyloid fibrils [29]. This effect is not limited to amyloidogenic proteins, defined in this text as those capable to form amyloid deposits in vivo. It has also been shown that the stressing conditions also promote the aggregation in the form of amyloid-like fibrils of a number of proteins not associated with the in vivo amyloid deposition [30]. A good example is the myoglobin, an α-helix rich protein that, under appropriated destabilizing conditions, can aggregate in the form of amyloid-like fibrils in the lab [31] Fig. 1h. These observations indicate that the loss of the native state is an essential step required for aggregation, because the native state is, in essence, non-amyloidogenic. The accumulation of non-native intermediates under conditions favoring fibrillogenesis has been demonstrated in several protein systems [29, 32–39]. Given that their presence generally correlates with faster kinetics of fibril formation, it is believed that they are key components of the aggregation pathway [38, 40–43]. Thus, much effort has been invested to determine the structure of such non-native species, using many different methodological approaches. Despite that, only in a limited number of studies it has been possible to obtain structural data of relatively high resolution [36, 44]. This is because the partially folded intermediates, whether they are exclusively part of the folding pathway or are also involved in the aggregation reaction, tend to be of transient existence and usually represent a minor fraction of the total population [45–49]. Despite this, it is generally accepted that protein self-assembly into amyloid fibrils occurs through conformational changes, which may affect the entire molecule or be circumscribed to just a part of it, resulting in exposure of aggregation promoter regions on the molecular surface, which are normally buried in the native state [14, 44, 50–52]. This general model is consistent with the observed inverse correlation between thermodynamic stability and protein amyloidogenic potential, since the probability that a molecule adopts a non-native conformation increases to the extent that the native folding is less stable [29, 53, 54]. Interestingly, it has been shown that in a relatively large number of proteins there are at least indirect evidences that point to the presence of one or more partially folded intermediates in the folding pathway [38, 55, 56]. However, only a very limited set of proteins have the capacity to form amyloids, even if those that only form such aggregates in the laboratory are taken into account. Thus, the capacity to populate a non-native intermediate as part of the folding/unfolding pathway or even off-pathway seems not to be enough for a protein to acquire the ability of self-assembly into the highly ordered structure of the amyloid fibril. Then, what structural factors make the proteins competent to aggregate in the form of amyloid fibrils?

Structural Bases of the Amyloid Fibril Formation

The causative implication of the amyloid in high impact human diseases, such as type 2 diabetes mellitus, Alzheimer's, Parkinson's, and prion diseases, made their structural characterization an attractive goal for various research groups with expertise in the structural study of biological macromolecules. However, the supramolecular, insoluble, and non-crystalline character of the amyloid fibril has complicated the task of determining its structure at the atomic level. This limitation is due to the fact that traditional methods used to determine such structures; in particular X-ray crystallography and NMR spectroscopy; are not directly applicable to molecules in

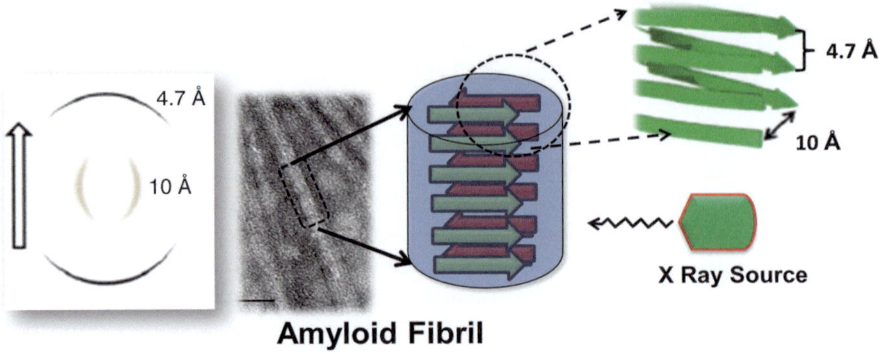

Fig. 3 Schematic representation of the cross-β X-ray diffraction pattern typically produced by the amyloid fibrils. The cross-β pattern is consistent with a core of the fibrils being formed by extensive β sheets arranged along to the longitudinal axis of the fibril, while the β strands forming them are arranged perpendicular to this. The reflections at 4.7 and 10 Å, which are typical of the cross-β diffraction pattern, reflect, respectively, the inter-strand and the inter-sheet spacing characterizing the cross-beta structure. The *white arrow* in the left corner indicates the relative orientation of the longitudinal axis of the fibrils regarding to the position of the pattern reflections

the aggregated state. One of the strategies that have been followed is the application of multiple biochemical and spectroscopic techniques in the analysis of the aggregates [57–62]. Such approach has generated relevant, although yet fragmented, information regarding the structural bases of amyloid fibril formation, leading to a better understanding of its biological properties [63–66].

At the end of the 1960s and the beginning of the 1970s of the past century, the first studies of X-ray diffraction of amyloid fibrils were reported [67]. These studies showed that amyloid fibrils generate a common diffraction pattern, composed of a pair of perpendicular reflections, located approximately at 4.7 and 10 Å, in the meridional and equatorial directions, respectively [67–69]. This diffraction pattern, referred to as "cross-β," is consistent with a fibrillar nucleus formed by extensive β-sheets that are positioned parallel to the longitudinal axis of the fibril, while the β-strands are perpendicular to this same axis [67–70] (Fig. 3).

More recently, in Dr. David Eisenberg's laboratory, a novel experimental approach was developed based on systematic analysis, by X-ray diffraction, of fibrocrystals formed by short fibrillogenic peptides [65, 71–73]. They demonstrated that, in the crystalline array, the peptides adopt an extended β-conformation, organized so that they form β-sheets that extend along to the longitudinal axis of the fibrocrystal. The β-sheets are stacked as the cross-β model predicts. It was also found that the side chains of the residues oriented towards the interface between the β-sheets interdigitate very complementarily, giving way to a compact structure that is referred to as an "anhydrous steric zipper" [73]. Based on these findings, it was proposed that the anhydrous steric zipper is the structural motif that characterizes the core of amyloid fibrils [73–75]. Since such a structure would require a high side chain complementarity of both charge and shape, and this is a function of the sequence, it can be deduced that the regions of the protein that form the fibril core must have specific patterns in their sequence. Diverse theoretical and experimental strategies have been developed to identify the sequence patterns suitable for amyloid formation. These efforts have given way for the development of a number of computational algorithms that allow the prediction, with a reasonable margin of error, of amyloid-prone sequences in proteins and peptides [76]. Based on studies about the correlation between sequence pattern and amyloidogenic potential of peptides and proteins, the "amyloid stretch hypothesis" was postulated, which proposes that the amyloidogenicity of a protein is localized in discrete segments of its primary sequence, characterized by specific sequence patterns [77, 78]. But, how frequent are amyloid-prone segments in proteins? David Eisenberg and coworkers searched for segments with high fibrillation propensity in all annotated ORFs in the *Escherichia coli*, *Saccharomyces cerevisiae*, and *Homo sapiens* genomes using the 3D profile method. They estimated that almost all ORFs contain at least one segment with high fibrillogenic propensity [75]. Such prediction is striking given the limited set of proteins that are known to form amyloid fibrils both *in vivo*, as well as in the laboratory. The authors conclude that there must be mechanisms that protect proteins from fibrillation. As the authors also state, in addition to containing segments with a sequence pattern that permits the formation of a steric zipper, the assembly of a protein into fibrils requires a sufficient conformational freedom of the self-complementary segments to interact with other molecules [75]. In globular proteins, this implies some degree of conformational reorganization, an event that will depend on how stable is the molecule [79, 80]. Structural rearrangement would also be required for weakening the protective anti-aggregation structural motifs that proteins have incorporated through evolution [81].

Thus, the hypotheses of the non-native intermediates and the amyloid stretch hypothesis, represent complementary views of the same phenomenon. They point to two key aspects of the general mechanism of amyloid fibril assembly, whose elucidation is required for a precise description of the amyloid aggregation mechanism of a globular protein: (1) The distribution and nature of amyloid-prone sequences and (2) The extension and characteristics of the conformational changes that expose those sequences—and probably others—to the intermolecular contacts.

Immunoglobulin Light Chain (AL) Amyloidosis

Immunoglobulin light chains are ~214 residue-long proteins that fold in two domains, one of which is variable (V_L) and the other constant (C_L), both with a β-sandwich structure, typical of the immunoglobulins (Fig. 4) [82].

Apart from their known role in immune response, being part of antibodies, light chains have a long history as tumoral biomarkers. Known as Bence Jones Proteins (BJP), which are free monoclonal light chains excreted in the urine, they were the first marker used for the diagnosis of multiple myeloma and other hematological malignancies [83]. Light chains were also the first amyloid protein identified

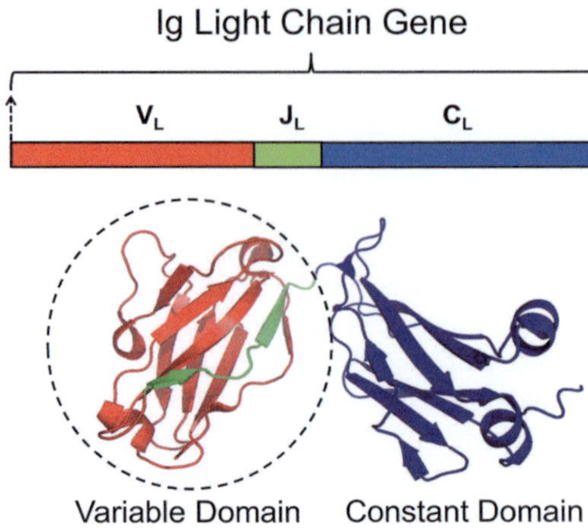

Fig. 4 Correlation between the molecular genetics and the 3D structure of the human immunoglobulin light chains. The functional gene of an individual light chain results from the recombination of three different genetic elements, the variable (V_L), joining (J_L) and constant (C_L) gene segments that are dispersed in the genome. The light chains fold into two domains, the variable, encoded by the V_L and J_L gene segments, and the constant, encoded by the C_L gene segment. The folding of both domains is characterized by the β-sandwich motif, typical of the immunoglobulins. The correspondence between the gene segment and the region of the protein that it encodes is indicated with the same color. The *cartoon representation* corresponds to the crystallographic structure of the λ6 light chain of an anti-hepcidin antibody (PDB ID 3H0T)

through amino acid sequencing. In 1970 Glenner and collaborators demonstrated that the fibrillar component of amyloid deposits, from a patient diagnosed with plasma cell discrasia and amyloidosis, was a κ-type light chain [84]. This finding showed that, under certain circumstances, light chains can form amyloid deposits, causing primary systemic amyloidosis, also known as AL amyloidosis. This disease is the systemic form of amyloidosis most commonly diagnosed in western countries and is characterized by the clonal expansion of plasma cells that invades poorly to moderately the bone marrow. The clonal expansion is accompanied by the excessive production of a light chain—which is by nature monoclonal—that is secreted in its free state, that is, not associated to the heavy chain in the context of a functional antibody [85]. This last condition is necessary for the amyloid deposition to occur, but it is not enough. Experimental evidence indicates that amyloid deposition of light chains is promoted by various factors, some dependent on the protein sequence, and others determined by the tissue microenvironment where aggregation occurs [86, 87]. Notably, while the intact monoclonal light chain may be part of the deposits, the most abundant components are usually fragments of the protein comprising the V_L or this domain plus a variable portion of the C_L [88]. It is yet to be established if proteolysis happens prior to aggregation or if it is a post-fibrillogenesis event [89]. Establishing the temporal relation between proteolysis and aggregation is of great significance for understanding the *in vivo* mechanism of AL deposit formation. Although it has been reported that the C_L can form amyloid deposits in certain conditions [90, 91], *in vitro* studies as well as those performed in animal models, indicate that the amyloidogenic potential of light chains is determined by the V_L sequence [92]. On the other hand, it has been observed that the kinetics of the *in vitro* aggregation of the isolated V_L is significantly faster than that of the intact light chain [93]. The slower aggregation kinetics of the whole light chain, compared to the isolated V_L, could be a consequence of the modulatory effects of the V_L-C_L interdomain contacts on the thermodynamics and/or kinetics of folding/unfolding of the protein [94]. An alternative explanation, not excluding the one just mentioned, is the inhibition of aggregation by steric hindrance, or by the mere increase in size of the molecule, because an inverse correlation between the molecular mass and the propensity of proteins to form amyloid fibrils has been observed [20, 21].

Among the tissue factors that can modulate light chain aggregation kinetics and fibril properties, glycosaminoglycans (GAGs) are the most studied extracellular components. As observed in other forms of amyloid deposits, fibrils extracted from tissue deposits of AL patients are invariably associated with GAGs, being heparan sulfate the predominant compound [95, 96]. It has been shown that some GAGs can interact stronger with the amyloidogenic free light chains than with the non amyloidogenic ones [97], promoting the aggregation of the former and in some cases stabilizing oligomeric species [98–100]. Recently, Marina Ramirez-Alvarado and coworkers reported that certain GAGs can modulate the kinetics of light chain fibril formation, an effect that seems to be associated with the natural propensity of each AL protein to form fibrils. Heparan sulfate showed a variable effect, as it accelerated fibril formation of some of the light chains, and delayed or exerted no effect on the fibrillogenesis of the others. In contrast, chondroitin sulfate A showed for all proteins a strong fibril formation inhibition. The authors propose that heparan sulfate facilitates the formation of transient amyloidogenic conformations of AL light chains, hereby promoting amyloid formation, whereas chondroitin sulfate A kinetically traps partially unfolded intermediates, inhibiting by this way the fibril elongation [101].

Although the contribution of tissue-dependent factors seems to be important, the evidences point to the physicochemical properties of the monoclonal light chain as the main determinant of its potential to aggregate into amyloid [86]. Similarly to what has been observed in other amyloid precursors [102–105], the propensity of the light chains to aggregate as amyloid correlates inversely with their thermodynamic stability [106–110]. The physicochemical properties of a protein are a function of its sequence. In the case of light chains, it should be mentioned that they are highly heterogeneous in sequence; something that makes them unique among amyloid precursor proteins [82]. Such sequence heterogeneity is linked to their recognition function as part of the antibodies and results from the mechanisms for the antibody repertoire diversification. Those mechanisms are the $V_L(D)J_L$ gene segments rearrangement, targeted deamination of cytosines, somatic hypermutation, gene conversion, and class-switch recombination [82]. There are strong evidences that both, the identity of the V_L gene segment that was used in the rearrangement, as well as the changes introduced in the germline sequences by the somatic hypermutation, contribute critically to the propensity of a light chain to deposit as amyloid *in vivo* [86, 87].

Somatic Mutation Contribution to Light Chain Amyloidogenesis

The somatic hypermutation of the antibody is an antigen-driven process that plays a fundamental role in the diversification of the humoral immune response, contributing to the increase in the recognition affinity of antibodies to the extent that the immune response against a specific antigen progresses [111]. In this regard, there are evidences that the amyloidogenic light chains derive from plasma cell clones that survived the process of antigen-driven selection [112]. In addition to the beneficial effect on antigen-antibody binding affinity, the somatic mutations can also exert a deleterious effects on the thermodynamic stability of the light chain, increasing the probability for the protein to adopt non-native transient states, prone to aggregation [113]. In studies published in the 1990s, it was shown that, in contrast to benign proteins, the amyloidogenic light chains are characterized by a mutation pattern that affects more frequently key structural positions of the V_L [108, 114]. Subsequent studies allowed the experimental validation of those findings and demonstrated a direct correlation between the destabilizing effect of mutations and the tendency of light chains to form amyloid fibril *in vitro* [109, 113, 115–117]. It has been found that mutations altering the structure and stability of the V_L-V_L dimer interface can promote light chains fibrillogenesis [106, 113]. In line with this finding, it was shown that the fibrillation kinetics of the κ4 recombinant V_L protein Len become faster by decreasing protein concentration or by adding urea, both factors promote the dissociation of the dimeric form of the protein [118]. Those findings are consistent with kinetic data showing that fibril elongation can be explained by a first-order kinetic model, meaning that the state that elongates the preformed fibrils is a monomer [119]. In conjunction, these findings support the concept that stabilizing a properly formed V_L-V_L dimer can be an effective strategy for inhibiting the light chain amyloidogenesis, as the light chain dimer is more stable and less fibrillogenic than the monomer [106].

In normal conditions, the destabilizing effect of somatic mutations on light chain folding can be effectively compensated by additional mutations that can be located in positions out of the antigen binding site [120]. Additionally, and not less relevant, the stabilization of a misfolding-prone light chain can result from the contacts that this protein establishes with the heavy chain in the quaternary structure of the antibody [86]. This is consistent with the idea that amyloid aggregation of light chains is partially a consequence of their abnormal secretion out of the molecular context—the whole antibody—in which they have functionally evolved [86].

Apart from their effect on the thermodynamic stability, somatic mutations can modify other light chain properties, such as solubility, surface charge distribution, and the propensity for β conformation of particular regions, which could also be important for the aggregation mechanism [121].

Contribution of the V_L Gene Segment

Around 90 % of the V_L domain sequence is encoded by the V_L gene segment while the rest is encoded by the J_L segment (Fig. 4) [82]. The repertoire of the V_L gene segments in humans is composed of around 70 elements, 30 being of λ type and 40

Table 1 Frequency of the five V_L gene segments most strongly associated to AL amyloidosis in the ALBase AL sequences

V_L gene segment	n[a]	%[b]
IGLV6-57 (λ6)	97	18.7 (25[c])
IGLV1-44 (λ1)	54	10.4
IGLV2-14 (λ2)	57	11.0
IGLV3-1 (λ3)	59	11.4
IGLK1-33 (κ1)	50	9.6
Sub-total	317	61
Total AL in ALBase	519[d] (388 ALλ and 131 ALκ)	100 (75 % ALλ and 25 % ALκ)

[a]Number of AL sequences in ALBase deriving from a particular V_L gene segment
[b]Percentage respect to the total of AL sequences in ALBase
[c]Percentage of ALλ sequences in ALBase which derive from the IGLV6-57 gene segment
[d]All the data presented in this table were taken from ALBase, at http://albase.bumc.bu.edu/aldb/

of κ type, these genes are located in chromosomes 22 and 2, respectively. Based on sequence similarity, the V_L gene segments have been divided into 10 λ (λ1 to λ10) and 4 κ (κ1 to κ4) major variable domain subgroups (http://www.vbase2.org/) [122]. The expression frequency of the V_L gene segments in the polyclonal B-cell repertoire varies significantly from one segment to another, with V_L genes physically located near locus J_L-C_L as the predominating group [123]. One significant finding is that the frequency with which each V_L gene segment is associated to amyloid deposition differs from its frequency of expression in the polyclonal B cell repertoire of peripheral blood and bone marrow [124, 125]. Just five V_L gene segments [*IGLV1-44* (*1c*), *IGLV2-14* (*2a2*), *IGLV3-1* (*3r*), *IGLV6-57* (*6a*), and *IGKV1-33* (*018/08*)] encode around 60 % of the amyloidogenic light chains whose sequence has been compiled in the ALBase (Table 1) [126]. These five segments represent less than 10 % of the human V_L repertoire [122]. Moreover, other V_L segments, frequently expressed in the normal polyclonal repertoire like *IGLV3-10* (*3p*), *IGLV3-25* (*3m*), and *IGLV1-40* (*1e*) are scarcely or not represented among the AL proteins in the ALBase [126]. The *6a* gene segment is the only member of the λ6 subgroup, which means that it encodes for all λ6 light chains. *6a* shows the strongest association with amyloidosis, since it encodes around 24 % of all ALλ proteins in the ALBase (Table 1) [126]. This figure contrasts with the low expression of *6a* in the polyclonal repertoire of the bone marrow plasma B-cells, which is of just 2 % [125]. It must be added that, with the exception of the Jto myeloma protein [109], all λ6 monoclonal light chains so far identified have been implicated in some form of amyloid deposition [127].

6aJL2 Protein as a Model of Amyloid Aggregation

A strong association of light chains belonging to the λ6 subgroup with amyloid aggregation has been confirmed in several studies [125, 127–130]. However, the molecular causes of this relationship have not been clearly established yet. As virtually all the monoclonal λ6 light chains so far identified were shown to be amyloidogenic, it has been suggested that some yet unestablished structural factors encoded by the germline *6a* gene segment confer to the λ6 proteins an abnormally high propensity to aggregate [127, 129, 130]. Given the well-documented inverse correlation between folding stability and amyloid aggregation, it was initially thought that the *6a* gene segment translates into an inherently unstable and misfolding-prone V_L domain. To test this hypothesis, we constructed a model protein named "6aJL2", which is a recombinant V_L domain with the predicted sequence of the germline gene segment *IGLV6-57* and the light chain join (J_L) gene segment *IGLJ2* (*JL2*) The *IGLJ2* gene segment was selected for this construct because it has been found to be the most frequently expressed J_L segment in the repertoire of ALλ proteins, as well as in the normal repertoire of polyclonal antibodies [125]. It was found that 6aJL2 forms amyloid fibrils under physiologically relevant conditions of temperature, pH and ionic strength. However, when this protein was compared to two naturally occurring λ6 V_L domains, the amyloidogenic Wil and the myeloma protein Jto, 6aJL2 resulted to be the most thermodynamically stable [131]. This finding suggested that the thermodynamic stability of the germline λ6 V_L protein does not explain by itself the high amyloidogenicity of the λ6 monoclonal light chains. Therefore, it was proposed that somatic mutations are also important [131].

Involvement of Partially Folded Intermediates in AL Fibril Assembly Mechanism

The thermodynamics and kinetics of 6aJL2 unfolding have been characterized by spectroscopic, hydrodynamic and calorimetric techniques. In the short time scale data were consistent with a two-state transition [132]. But, interestingly, after incubation for several hours at 37 °C, the presence of a partially unfolded off-pathway intermediate was revealed in the urea concentration interval of 1.5 to 3.5 M [132]. *In vitro* fibrillogenesis assays showed that the maximum growth rate for fibril formation and the shortest lag time were obtained at urea concentrations where the partially unfolded state was populated. Based on these findings, it was suggested that the detected non-native intermediate is a key component of the *in vitro* fibril formation pathway of 6aJL2 [132]. Its structure was not determined in detail, but the data generated in the study are consistent with a soluble oligomeric state. As the formation of the folding intermediate was larger around the midpoint of urea unfolding, where the native and unfolded states are equally populated and their rate of interconversion is the slowest, it was concluded that the acceleration of the fibrillogenesis of 6aJL2 is explained by the slow accumulation of an intermediate state that is prone to aggregate [132]. Others have also found evidence of the involvement of partially folded intermediates in the fibrillar aggregation of light chains. Fink and collaborators, while studying the kinetics of fibrillogenesis of the recombinant amyloidogenic V_L SMA (κ4), detected two partially folded conformational species that accumulate at low pH [133]. One of them, which accumulated between pH 4 and 6, was a relatively native-like intermediate characterized by significant tertiary structure changes, but with a conserved secondary structure. A second, relatively unfolded, but compact intermediate was observed at pH below 3. They found that the relatively unfolded intermediate readily formed amyloid fibrils, while the native-like one preferentially led to amorphous aggregates [133].

The two mentioned studies provided strong evidence of the involvement of partially folded species in the kinetic pathway of light chain amyloid aggregation. But, given the high structural heterogeneity of light chains—meaning many different molecular contexts in which aggregation can occur—further studies are needed to generate a more accurate view of the self-association mechanism(s) of light chains into the AL fibril. Transcendental issues, as the origin of the fibrillogenic non-native specie(s)—on- or off-pathway intermediates—and the extent of unfolding required for amyloid aggregation to occur, need to be studied thoroughly.

Sequence as a Driving Force of Light Chain Aggregation

Solomon and coworkers showed many years ago that Bence Jones proteins (BJP) purified from the urine of patients with AL amyloidosis or multiple myeloma (MM) can reproduce in mice the same type of renal deposits that they caused in the patients from which they were obtained [92, 134, 135]. Moreover, Sikkink, L. A. & Ramirez-Alvarado, M., evaluated the folding thermodynamics and aggregation properties of six BJP purified from the urine of patients with AL, light chain deposition disease (LCDD) and MM [136]. They found that AL proteins are the least thermodynamically stable of all BJP. In order to evaluate the aggregation of proteins in an energetically equivalent condition, the aggregation assays were performed at the midpoint of thermal unfolding, in which the half of the molecules are fully folded and half are fully unfolded. Notably, it was observed that only the AL proteins formed fibrils, while the LCDD and MM proteins formed, respectively, amorphous aggregates and large spherical species [136]. These findings can be interpreted as the manifestation of the modulating effect of putative sequence elements on the aggregation landscape of light chains. Such reasoning is in line with the central postulate of the "amyloid stretch hypothesis", which considers the amyloidogenesis of a protein as the manifestation of a dominant driving effect of short sequence stretches over the aggregation propensity of the entire molecule. Such amyloid-prone sequence stretches have been successfully identified in several amyloid-forming proteins [137–140], but almost nothing is known regarding their identity in light chains.

Predicting the Fibrillogenic Segments of the 6aJL2 Protein

We have used two web-based computational algorithms for predicting the amyloid-prone regions of 6aJL2 proteins. One of them is the 3D Profile Method, which is a structure-based algorithm developed in the lab of Dr. David Eisenberg, at the

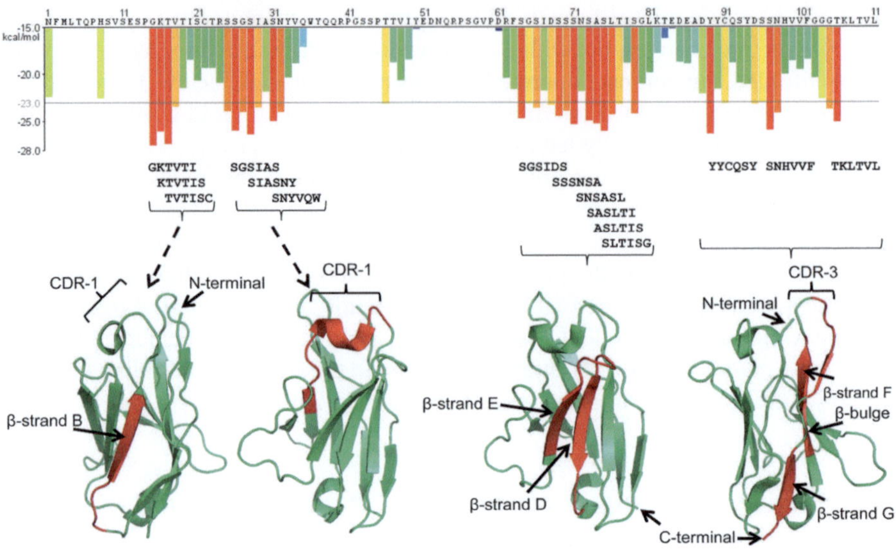

Fig. 5 Highly amyloid-prone hexapeptides predicted by the 3D Profile Method on the sequence of 6aJL2 protein. The image of the histogram representation was taken directly from the web page of the tool (http://services.mbi.ucla.edu/zipperdb/protein/23772). The *orange/red colored bars* represent segments with energy below the indicated energetic threshold of –23 kcal/mol (*gray line*), which are predicted to form fibrils. The sequences listed below the histogram correspond to the hexapeptides with a Rosetta energy of ≤–25 kcal/mol, which are represented by *red-colored* bars. As mentioned in the text, this conservative energy threshold results in the recovering of the 69 % of the positive sequences, with a false-positive error rate of 22 % [141]. In the figure bottom, the structural location of the predicted highly fibrillogenic hexapeptides is highlighted in *red* in the cartoon representations of the 6aJL2 proteins (PDB ID 2W0K)

University of California Los Angeles [141]. The 3D Profile Method was developed based on the crystal structure of the cross-β spine formed by the peptide GNNQQNY from the sup35 prion protein of *Saccharomyces cerevisiae* [71]. The window of searching through the protein sequence has a length of six residues; each six-residue stretch is mapped onto an ensemble of templates generated from the crystal structure of the peptide NNQQNY and the energy of each mapping is evaluated by using ROSETTADESIGN [141]. The lowest energy match for a given hexapeptide stretch to the template library is taken as the putative prediction. A threshold value of –23 kcal/mol is normally set in the algorithm for predicting the sequences with a high fibrillogenic propensity. The authors have reported that when a permissive threshold value of –19 kcal/mol is used, 100 % of the positive sequences are recovered, but with a false-positive error rate of 45 %. Instead, when the threshold value is set at –25.5 kcal/mol, the 69 % of the positive sequences are recovered, with a false-positive error rate of 22 % [141]. Searching onto the 6aJL2 sequence led to the identification of several hexapeptides with Rosetta energy lower than –23 kcal/mol. Several of them overlap into three aggregation-prone regions that locate, respectively, at the beta-strand B, the CDR1 and the segment of the FR3 formed by β-strands D and E and the loop linking them (Fig. 5).

The second algorithm used for analyzing 6aJL2 protein was AmylPred2, a consensus prediction algorithm of amyloidogenic sequences in globular proteins developed in the Laboratory of Professor Stavros J. Hamodrakas at the University of Athens [142]. AmylPred2 was built pursuing the goal of a more objective and accurate prediction of the amyloidogenic determinants in proteins. It combines 11 existing individual web tools for predicting aggregation-prone/amyloidogenic segments in protein from sequence alone. The consensus of these methods is defined as the hit overlap of 5 out of 11 methods [142]. The predictions on 6aJL2 sequence of each individual algorithm, as well as the consensus amyloid-prone sequences are shown in Fig. 6. As can be noted when comparing Figs. 5 and 6, the predictions generated by the two algorithms—3D Profile Method and Amyl-Pred 2—overlap in large extent to each other. Both algorithms recognized the β-strand B, the C-terminal half of the CDR-1, and the β-strands C and E and part of the β-strand G, as amyloid-prone segments. It should be highlighted that β-strands B, C and G are edges of β sheets of the sandwich, while the C-terminal half of the CDR1 loop folds as a distorted helix and is positioned on the top of the domain. From a structural point of view, the CDR1 loop is a β-arm connecting β-strands B y C.

Experimental Identification of the Amyloid-Prone Regions in 6aJL2 Protein: How Well Do the Predictions Match the Experimental Data?

In order to test the predictions of aggregation-prone sequences performed by the two algorithms described above, the aggregation behavior of a set of 32 different synthetic peptides, 31 hexapeptides and 1 decapeptide, was evaluated (del Pozo-Yauner, L., manuscript in preparation). The set included the 15 hexapeptides with a Rosetta energy ≥25 kcal/mol, predicted by the 3D Profile Method to be highly fibrillogenic (Fig. 5 and Table 2). It also encompassed ten hexapeptides with Rosetta energy < −23 kcal/mol and nine hexapeptides with energy between −23 and −25 kcal/mol. The decapeptide evaluated in this study has the 6aJL2 sequence from Ile30 to Gln37 ($_{30}$IASNYVQWYQ$_{37}$), being different in only one residue from the consensus sequence $_{30}$ASNYVQWYQ$_{37}$ generated by Amyl-Pred 2. Ile30 was added because it is present in the prediction of four out of the ten individual methods from which the consensus of Amyl-Pred 2 was built (Fig. 6). This includes the predictor Waltz, which has been optimized for predicting amyloidogenic sequences [144, 145].

The experimental methodology was as follows: Aggregation assays were performed at 250 μM peptide concentration, diluted in buffered phosphate saline pH 7.4 (PBS). Triplicate samples were added to 2 mL boil-proof plastic microtubes with conical bottom (Axygen Cat. No. MCT-200-C) and incubated at 37 °C for 24 h, with constant orbital agitation (1000 r.p.m.) in a Thermomixer Comfort (Eppendorff). Fibril formation was evaluated by the thioflavin T (ThT) assay and end-point samples were analyzed by electron microscopy (EM).

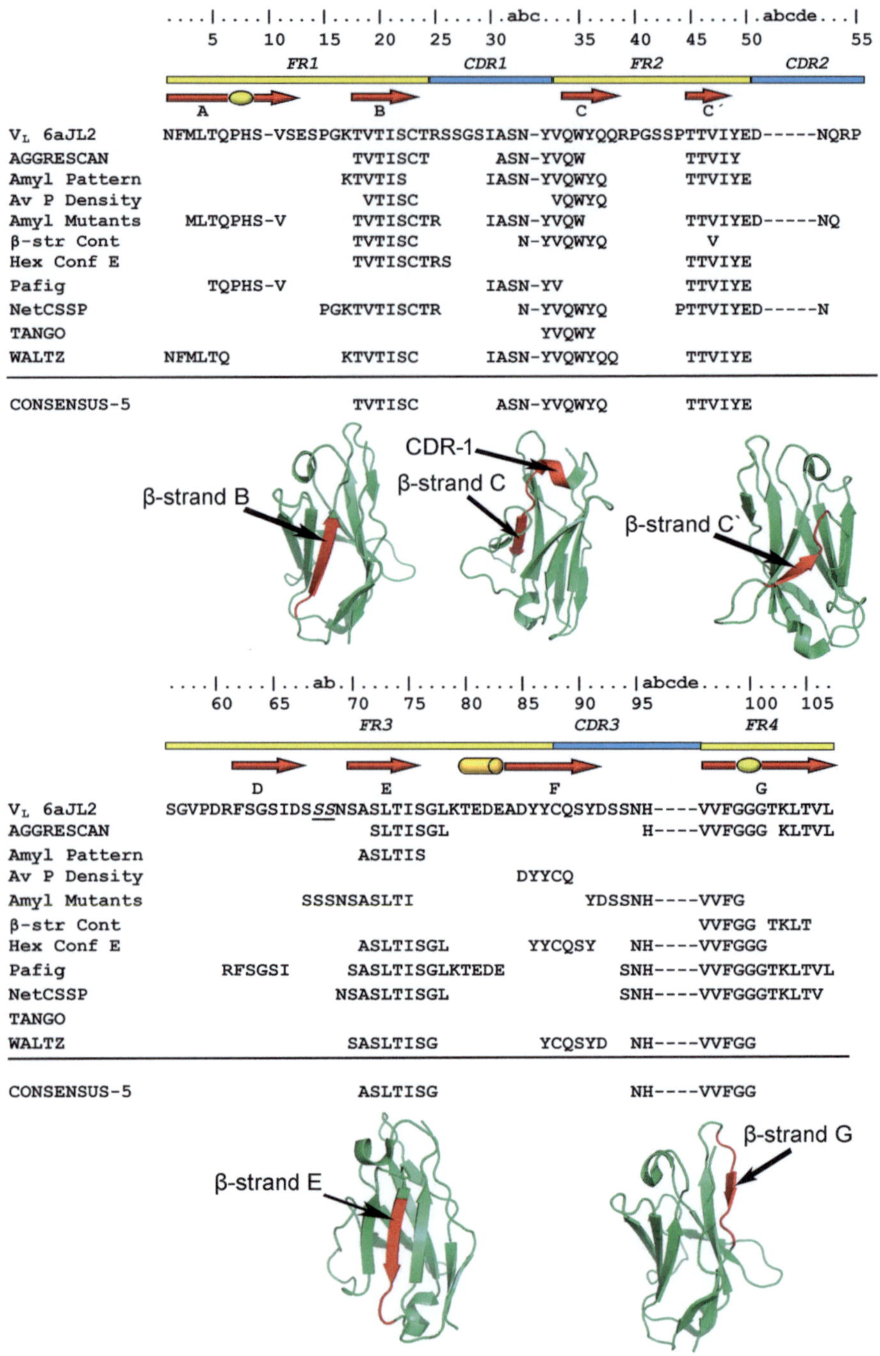

Fig. 6 Consensus sequences generated by the Amyl-Pred 2 method on the basis of the analysis of the 6aJL2 protein with 11 different algorithms, designed for predicting aggregation/fibrillogenic-prone sequences [142]. The individual prediction of each method is also shown. As the SecStr (Possible Conformational Switches) method gave a null result on the sequence of 6aJL2 protein,

Table 2 Set of 32 synthetic peptides whose aggregation behavior was evaluated for testing the predictions of the computational algorithms 3D profile method and Amyl-Pred 2, as described in the text

Sample	Molecule I.D.	Sequence (one letter code)	Rosetta energy[a] (kcal/mol)	ThT assay[b]	Electron microscopy[c]
1	L6(N1-Q6)	NFMLTQ	−22.3	–	–
2	L6(H8-S14)	HSVSES	−22.5	–	–
3	L6(G16-I21)	GKTVTI	−27.5	–	–
4	L6(K17-S22)	KTVTIS	−26.0	–	–
5	L6(T18-C23)	TVTISC	−27.3	–	–
6	L6(T20-R25)	TISCTR	−21.4	–	–
7	L6(S27-S31a)	SGSIAS	−25.9	–	–
8	L6(S29-Y32)	SIASNY	−26.3	–	–
9	L6(I30-V33)	IASNYV	−23.5	–	Fibrillar
10	L6(S31a-W35)	SNYVQW	−24.9	–	Microcrystals
11	L6(Y32-Q37)	YVQWYQ	−20.3	–	–
12	L6(T45-E50)	TTVIYE	−23.1	–	–
13	L6(S63-S68)	SGSIDS	−24.7	–	–
14	L6(S68-A71)	SSSNSA	−24.4	–	–
15	L6(S68b-L73)	SNSASL	−25.3	–	–
16	L6(N69-T74)	NSASLT	−21.8	–	–
17	L6(S70-I75)	SASLTI	−24.9	–	–
18	L6(A71-S76)	ASLTIS	−25.2	–	–
19	L6(S72-G77)	SLTISG	−26.0	–	–
20	L6(L73-L78)	LTISGL	−24.3	–	–
21	L6(T74-K79)	TISGLK	−23.2	–	–
22	L6(I75-T80)	ISGLKT	−18.8	–	–
23	L6(S76-E81)	SGLKTE	−24.2	–	–
24	L6(T80-D85)	TEDEAD	−16.2	–	–
25	L6(E81-Y86)	EDEADY	−14.5	–	–
26	L6(Y86-Y91)	YYCQSY	−26.3	–	–
27	L6(Y87-D92)	YCQSYD	−21.5	–	–
28	L6(S94-F98)	SNHVVF	−25.9	–	–
29	L6(N95-G99)	NHVVFG	−24.1	–	–
30	L6(H95a-G100)	HVVFGG	−20.0	–	–
31	L6(T102-L017)	TKLTVL	−25.1	–	–
32	L6(I30-Q37)	IASNYVQWYQ	N/A	+	Fibrillar and microcrystals

[a]Calculated by ROSETTADESIGN algorithm [141]
[b]Performed as described in del Pozo-Yauner et al. [19]
[c]Performed as described in del Pozo-Yauner et al. [19]

Fig. 6 (continued) it was not included in the figure. The regions of 6aJL2 protein with β-strands or helix conformation in the native state are indicated by *arrows* and *cylinder*, respectively. The *oval figures* in the first and last *arrows* represent, respectively, the sheet-switch motif characterizing the structure of the N-terminal segment, and the β-bulge centered at Gly100 in the β-strand G. The residue numbering and the location of the CDR/FR regions are according to Kabat et al. [143]. The two underlined SS represent two Ser residues considered, by structural alignment, inserted between positions 68 and 69 in the λ6 light chains

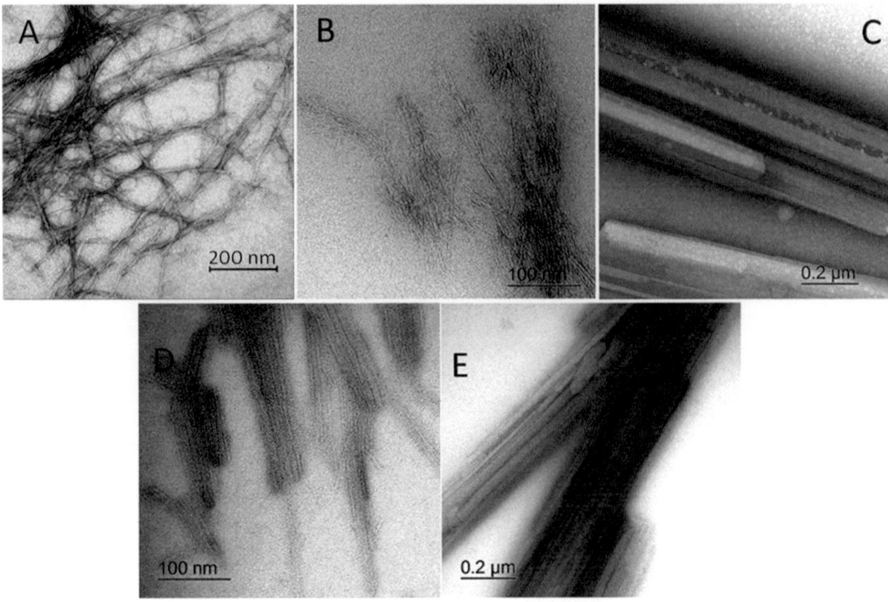

Fig. 7 Electron micrographs of the aggregates formed by (**a**) the rV_L protein 6aJL2, and the synthetic peptides (**b**) $_{31a}$SNYVQW$_{35}$ (sample 10), (**c**) $_{32}$YVQWYQ$_{37}$ (sample 11) and (**d**, **e**) $_{30}$IASNYVQWYQ$_{37}$ (sample 32), whose sequences derive from the 6aJL2 protein. The subscripts indicate the numbering of the N- and C-terminal residues in the primary structure of 6aJL2 protein

After ~2 h of incubation, samples 10 ($_{31a}$SNYVQW$_{35}$) and 32 ($_{30}$IASNYVQWYQ$_{37}$) turned opalescent, while a similar change in appearance was observed in sample 11 ($_{32}$YVQWYQ$_{37}$) only after overnight incubation. The other samples kept clear after 24 h of incubation. The ThT assay demonstrated that only sample 32 contained aggregates with the capacity to fluoresce in presence of ThT. E.M. analysis showed that samples 10 (Fig. 7b) and 32 (Fig. 7d) contained abundant fibrillar aggregates that tend to associate in bundles, while sample 11 (Fig. 7c), as well as sample 32 (Fig. 7e), contained aggregates with the appearance of microcrystals (del Pozo-Yauner, L. manuscript in preparation). The fibrils of 6aJL2 protein are long and appear to be formed by two protofibrils that braid each other (Fig. 7a). Moreover, the fibrils of samples 10 and 32 are thinner and a little shorter than the fibrils of 6aJL2 protein, without an evident sub-structural organization. It should be noted that the peptides 10 ($_{31a}$SNYVQW$_{35}$) and 11 ($_{32}$YVQWYQ$_{37}$) partially overlap each other and both are contained in the sequence of peptide 32 ($_{30}$IASNYVQWYQ$_{37}$). It has been shown that some peptides can form microcrystals or fibrils, or both, in a manner that depends of the sequence and the environmental conditions [73]. It is thought that, in the fibrillar state, the dominant intermolecular contacts result in the arrangement of the molecules in the form of long intermolecular β sheets, stacked in layers and oriented along to the longitudinal axis of the fibrils [64, 69]. In the case

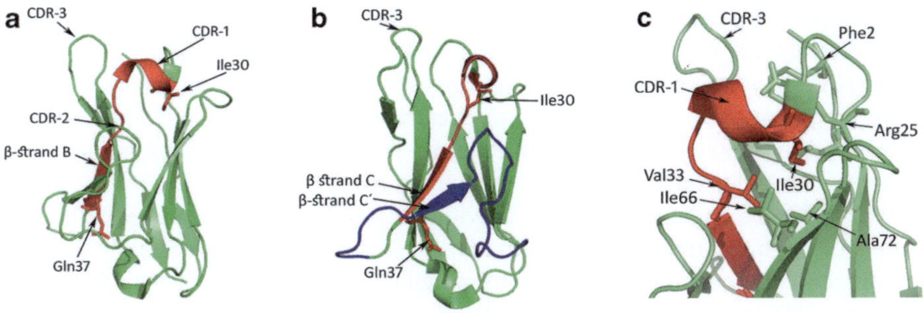

Fig. 8 (**a**) Localization of the fibrillogenic region Ile30-Gln37 (colored in *red*) in the 3D structure of the rV_L protein 6aJL2 . (**b**) Structural relation of the β strand C, which is part of the fibrillogenic region shown in (**a**), with the segment spanning from Pro40 to Asp60 (colored in *blue*). (**c**) Structural relationship of Ile30 with the key residues determining the peptide backbone conformation of the CDR1 in the λ6 light chains [146]

of the microcrystal, the peptides also associate to form stacked β sheets, similarly to the fibrils, but, an additional set of regular contacts stabilizes the 3D crystalline lattice [72, 73]. Thus, the ability of the peptides 10 and 11 to aggregate differently in similar environmental conditions most probably reflects the modulating effect of the sequence over the contacts and forces that lead to, and stabilize, the aggregated state. The capacity of the peptide 32 to form both type of aggregates, fibrils and microcrystals, suggests that different sequence elements, present respectively in peptides 10 and 11, determine such behavior.

In conclusion, our experiments showed that the peptide 32 ($_{30}$IASNYVQWYQ$_{37}$) forms amyloid-like fibrils in physiological conditions of temperature, pH and ionic strength. This finding indicates that the corresponding region of protein 6aJL2 contains at least one amyloid-prone sequence. If this region plays a role in the mechanism of amyloid aggregation of the 6aJL2 protein is at present unknown. Beyond this, some implications for the mechanism of amyloidogenesis of 6aJL2 proteins can be anticipated from a detailed analysis of its crystallographic structure. In the native state, the fibrillogenic region spanning from Ile30 to Gln37 folds as the C-terminal half of the Complementary Determining Region 1 (CDR1) and the β strand B (Fig. 8). The CDR1 in the 6aJL2 protein adopts a distinctive backbone conformation, characterized by a distorted helix from Ile30 to Asn31b. The comparative structural analysis led to classify the CDR1 conformation of the λ6 light chains as a new type of canonical structure, distinctive of this family of proteins [146]. The side chain of Ile30 is placed deep inside a nonpolar pocket formed by several residues that are part of the hydrophobic core of the domain. By means of these contacts, it is thought that Ile30 contributes critically to the stability of the CDR1 structure [146]. Moreover, β strand B is located at the edge of one of the two sheets forming the domain β sandwich (Fig. 8). In agreement with the findings of J.S. Richardson & D.C. Richardson, regarding the edge-protection strategies used by β structure-rich natural proteins [81], β strand B is protected by the long segment spanning from

Pro40 to Asp60. Thus, based on these structural features, it can be inferred that some sort of structural rearrangement of the helical segment of the CDR1, including changes in the relative positioning of the key residue Ile30, should occur as part of the aggregation mechanism. The putative rearrangements should extend to the segment Pro40-Asp60, a step probably required for exposing the aggregation-prone sequence located at the β strand C to the intermolecular contacts.

The evidence indicates that the amyloidogenesis of the immunoglobulin LC occurs through a complex mechanism that implies some degree of conformational rearrangement. Several factors seem to modulate the propensity of the LC to self-assembly into amyloid fibrils. Among the most important are the folding stability, the presence of pro-amyloidogenic sequences, the composition of the extracellular matrix, the plasma concentration of the monoclonal LC itself, and the degree of conservation of the renal function. Differing from the other amyloid precursors, the LC are very heterogeneous in sequence, a characteristic that is inherent to their recognition function as part of the antibodies. Such structural heterogeneity has complicated the task of unravel the mechanism of the AL formation. In this chapter, we have summarized the most recent findings regarding the role of the partially folded intermediates, as well as the influence of other factors, as the components of the extracellular matrix, in the mechanism of amyloidogenesis. Also, we have discussed the available data regarding the contribution of the somatic mutations and the pro-amyloidogenic regions to the aggregation propensity of the LC. Despite the advances obtained in the studies referred in this document, many key questions regarding the LC aggregation mechanism remain unanswered. Thus, further researches are needed to understand why and how some light chains form amyloid fibrils in vivo while others do not.

Acknowledgment This work was supported in part by grants from the Consejo Nacional de Ciencia y Tecnología (No. 169659) to L. del Pozo-Yauner.

References

1. Andras P, Andras C (2005) The origins of life—the 'protein interaction world' hypothesis: protein interactions were the first form of self-reproducing life and nucleic acids evolved later as memory molecules. Med Hypotheses 64(4):678–688. doi:10.1016/j.mehy.2004.11.029
2. Pirlet K, Arthur-Goettig A (1999) Maintaining life and health by natural selection of protein molecules. J Theor Biol 201(1):75–85. doi:10.1006/jtbi.1999.1015
3. Hingorani KS, Gierasch LM (2014) Comparing protein folding in vitro and in vivo: foldability meets the fitness challenge. Curr Opin Struct Biol 24:81–90. doi:10.1016/j.sbi.2013.11.007
4. Ferris SP, Kodali VK, Kaufman RJ (2014) Glycoprotein folding and quality-control mechanisms in protein-folding diseases. Dis Model Mech 7(3):331–341. doi:10.1242/dmm.014589
5. Dobson CM (2003) Protein folding and misfolding. Nature 426(6968):884–890. doi:10.1038/nature02261
6. Hipp MS, Park SH, Hartl FU (2014) Proteostasis impairment in protein-misfolding and -aggregation diseases. Trends Cell Biol 24(9):506–514. doi:10.1016/j.tcb.2014.05.003
7. Stefani M, Dobson CM (2003) Protein aggregation and aggregate toxicity: new insights into protein folding, misfolding diseases and biological evolution. J Mol Med 81(11):678–699. doi:10.1007/s00109-003-0464-5

8. Papsdorf K, Richter K (2014) Protein folding, misfolding and quality control: the role of molecular chaperones. Essays Biochem 56:53–68. doi:10.1042/bse0560053
9. Wyatt AR, Yerbury JJ, Ecroyd H, Wilson MR (2013) Extracellular chaperones and proteostasis. Annu Rev Biochem 82:295–322. doi:10.1146/annurev-biochem-072711-163904
10. Gong H, Yang X, Zhao Y, Petersen RB, Liu X, Liu Y, Huang K (2014) Amyloidogenicity of p53: a hidden link between protein misfolding and cancer. Curr Protein Pept Sci 16(2):135–146
11. Nagaraj NS, Singh OV, Merchant NB (2010) Proteomics: a strategy to understand the novel targets in protein misfolding and cancer therapy. Expert Rev Proteomics 7(4):613–623. doi:10.1586/epr.10.70
12. Ashraf GM, Greig NH, Khan TA, Hassan I, Tabrez S, Shakil S, Sheikh IA, Zaidi SK, Akram M, Jabir NR, Firoz CK, Naeem A, Alhazza IM, Damanhouri GA, Kamal MA (2014) Protein misfolding and aggregation in Alzheimer's disease and type 2 diabetes mellitus. CNS Neurol Disord Drug Targets 13(7):1280–1293
13. Negahdar M, Aukrust I, Molnes J, Solheim MH, Johansson BB, Sagen JV, Dahl-Jorgensen K, Kulkarni RN, Sovik O, Flatmark T, Njolstad PR, Bjorkhaug L (2014) GCK-MODY diabetes as a protein misfolding disease: the mutation R275C promotes protein misfolding, self-association and cellular degradation. Mol Cell Endocrinol 382(1):55–65. doi:10.1016/j.mce.2013.08.020
14. Iram A, Naeem A (2014) Protein folding, misfolding, aggregation and their implications in human diseases: discovering therapeutic ways to amyloid-associated diseases. Cell Biochem Biophys 70(1):51–61. doi:10.1007/s12013-014-9904-9
15. Knowles TP, Vendruscolo M, Dobson CM (2014) The amyloid state and its association with protein misfolding diseases. Nat Rev Mol Cell Biol 15(6):384–396. doi:10.1038/nrm3810
16. Sideras K, Gertz MA (2009) Amyloidosis. Adv Clin Chem 47:1–44
17. Buxbaum JN, Linke RP (2012) A molecular history of the amyloidoses. J Mol Biol 421(2–3):142–159. doi:10.1016/j.jmb.2012.01.024
18. Hazenberg BP (2013) Amyloidosis: a clinical overview. Rheum Dis Clin North Am 39(2):323–345. doi:10.1016/j.rdc.2013.02.012
19. del Pozo-Yauner L, Wall JS, Gonzalez Andrade M, Sanchez-Lopez R, Rodriguez-Ambriz SL, Perez Carreon JI, Ochoa-Leyva A, Fernandez-Velasco DA (2014) The N-terminal strand modulates immunoglobulin light chain fibrillogenesis. Biochem Biophys Res Commun 443(2):495–499. doi:10.1016/j.bbrc.2013.11.123
20. Baldwin AJ, Knowles TP, Tartaglia GG, Fitzpatrick AW, Devlin GL, Shammas SL, Waudby CA, Mossuto MF, Meehan S, Gras SL, Christodoulou J, Anthony-Cahill SJ, Barker PD, Vendruscolo M, Dobson CM (2011) Metastability of native proteins and the phenomenon of amyloid formation. J Am Chem Soc 133(36):14160–14163. doi:10.1021/ja2017703
21. Thirumalai D, Reddy G (2011) Protein thermodynamics: are native proteins metastable? Nat Chem 3(12):910–911. doi:10.1038/nchem.1207
22. Anfinsen CB (1973) Principles that govern the folding of protein chains. Science 181(4096):223–230
23. Colon W, Lai Z, McCutchen SL, Miroy GJ, Strang C, Kelly JW (1996) FAP mutations destabilize transthyretin facilitating conformational changes required for amyloid formation. Ciba Found Symp 199:228–238; discussion 239–242
24. Maury CP, Nurmiaho-Lassila EL, Rossi H (1994) Amyloid fibril formation in gelsolin-derived amyloidosis. Definition of the amyloidogenic region and evidence of accelerated amyloid formation of mutant Asn-187 and Tyr-187 gelsolin peptides. Lab Invest 70(4):558–564
25. Martsev SP, Dubnovitsky AP, Vlasov AP, Hoshino M, Hasegawa K, Naiki H, Goto Y (2002) Amyloid fibril formation of the mouse V(L) domain at acidic pH. Biochemistry 41(10):3389–3395
26. Mishra R, Sorgjerd K, Nystrom S, Nordigarden A, Yu YC, Hammarstrom P (2007) Lysozyme amyloidogenesis is accelerated by specific nicking and fragmentation but decelerated by

intact protein binding and conversion. J Mol Biol 366(3):1029–1044. doi:10.1016/j.jmb.2006.11.084
27. Jansen R, Dzwolak W, Winter R (2005) Amyloidogenic self-assembly of insulin aggregates probed by high resolution atomic force microscopy. Biophys J 88(2):1344–1353. doi:10.1529/biophysj.104.048843
28. Wiseman RL, Green NS, Kelly JW (2005) Kinetic stabilization of an oligomeric protein under physiological conditions demonstrated by a lack of subunit exchange: implications for transthyretin amyloidosis. Biochemistry 44(25):9265–9274. doi:10.1021/bi050352o
29. Serebryany E, King JA (2014) The betagamma-crystallins: native state stability and pathways to aggregation. Prog Biophys Mol Biol 115(1):32–41. doi:10.1016/j.pbiomolbio.2014.05.002
30. Hamada D, Dobson CM (2002) A kinetic study of beta-lactoglobulin amyloid fibril formation promoted by urea. Protein Sci 11(10):2417–2426. doi:10.1110/ps.0217702
31. Fandrich M, Forge V, Buder K, Kittler M, Dobson CM, Diekmann S (2003) Myoglobin forms amyloid fibrils by association of unfolded polypeptide segments. Proc Natl Acad Sci U S A 100(26):15463–15468. doi:10.1073/pnas.0303758100
32. Armen RS, Alonso DO, Daggett V (2004) Anatomy of an amyloidogenic intermediate: conversion of beta-sheet to alpha-sheet structure in transthyretin at acidic pH. Structure 12(10):1847–1863. doi:10.1016/j.str.2004.08.005
33. Ferrao-Gonzales AD, Souto SO, Silva JL, Foguel D (2000) The preaggregated state of an amyloidogenic protein: hydrostatic pressure converts native transthyretin into the amyloidogenic state. Proc Natl Acad Sci U S A 97(12):6445–6450
34. Gupta P, Deep S (2014) Intermediate conformation between native beta-sheet and non-native alpha-helix is a precursor of trifluoroethanol-induced aggregation of human carbonic anhydrase-II. Biochem Biophys Res Commun 449(1):126–131. doi:10.1016/j.bbrc.2014.04.160
35. Jahn TR, Parker MJ, Homans SW, Radford SE (2006) Amyloid formation under physiological conditions proceeds via a native-like folding intermediate. Nat Struct Mol Biol 13(3):195–201. doi:10.1038/nsmb1058
36. Kameda A, Hoshino M, Higurashi T, Takahashi S, Naiki H, Goto Y (2005) Nuclear magnetic resonance characterization of the refolding intermediate of beta2-microglobulin trapped by non-native prolyl peptide bond. J Mol Biol 348(2):383–397. doi:10.1016/j.jmb.2005.02.050
37. Pallares I, Vendrell J, Aviles FX, Ventura S (2004) Amyloid fibril formation by a partially structured intermediate state of alpha-chymotrypsin. J Mol Biol 342(1):321–331. doi:10.1016/j.jmb.2004.06.089
38. Santucci R, Sinibaldi F, Fiorucci L (2008) Protein folding, unfolding and misfolding: role played by intermediate states. Mini Rev Med Chem 8(1):57–62
39. Vanderhaegen S, Fislage M, Domanska K, Versees W, Pardon E, Bellotti V, Steyaert J (2013) Structure of an early native-like intermediate of beta2-microglobulin amyloidogenesis. Protein Sci 22(10):1349–1357. doi:10.1002/pro.2321
40. Agocs G, Szabo BT, Kohler G, Osvath S (2012) Comparing the folding and misfolding energy landscapes of phosphoglycerate kinase. Biophys J 102(12):2828–2834. doi:10.1016/j.bpj.2012.05.006
41. Harrison RS, Sharpe PC, Singh Y, Fairlie DP (2007) Amyloid peptides and proteins in review. Rev Physiol Biochem Pharmacol 159:1–77. doi:10.1007/112_2007_0701
42. Murphy RM (2007) Kinetics of amyloid formation and membrane interaction with amyloidogenic proteins. Biochim Biophys Acta 1768(8):1923–1934. doi:10.1016/j.bbamem.2006.12.014
43. Sarkar N, Dubey VK (2013) Exploring critical determinants of protein amyloidogenesis: a review. J Pept Sci 19(9):529–536. doi:10.1002/psc.2539
44. Neudecker P, Robustelli P, Cavalli A, Walsh P, Lundstrom P, Zarrine-Afsar A, Sharpe S, Vendruscolo M, Kay LE (2012) Structure of an intermediate state in protein folding and aggregation. Science 336(6079):362–366. doi:10.1126/science.1214203
45. Ferreira ST, Chapeaurouge A, De Felice FG (2005) Stabilization of partially folded states in protein folding/misfolding transitions by hydrostatic pressure. Braz J Med Biol Res 38(8):1215–1222. doi:/S0100-879X2005000800009

46. Ferreira ST, De Felice FG, Chapeaurouge A (2006) Metastable, partially folded states in the productive folding and in the misfolding and amyloid aggregation of proteins. Cell Biochem Biophys 44(3):539–548. doi:10.1385/CBB:44:3:539
47. Jenkins DC, Sylvester ID, Pinheiro TJ (2008) The elusive intermediate on the folding pathway of the prion protein. FEBS J 275(6):1323–1335. doi:10.1111/j.1742-4658.2008.06293.x
48. Korzhnev DM, Religa TL, Banachewicz W, Fersht AR, Kay LE (2010) A transient and low-populated protein-folding intermediate at atomic resolution. Science 329(5997):1312–1316. doi:10.1126/science.1191723
49. Neira JL (2013) NMR as a tool to identify and characterize protein folding intermediates. Arch Biochem Biophys 531(1–2):90–99. doi:10.1016/j.abb.2012.09.003
50. Eichner T, Radford SE (2009) A generic mechanism of beta2-microglobulin amyloid assembly at neutral pH involving a specific proline switch. J Mol Biol 386(5):1312–1326
51. Guo Z, Eisenberg D (2007) The mechanism of the amyloidogenic conversion of T7 endonuclease I. J Biol Chem 282(20):14968–14974. doi:10.1074/jbc.M609514200
52. Nelson R, Eisenberg D (2006) Structural models of amyloid-like fibrils. Adv Protein Chem 73:235–282. doi:10.1016/S0065-3233(06)73008-X
53. Foss TR, Kelker MS, Wiseman RL, Wilson IA, Kelly JW (2005) Kinetic stabilization of the native state by protein engineering: implications for inhibition of transthyretin amyloidogenesis. J Mol Biol 347(4):841–854. doi:10.1016/j.jmb.2005.01.050
54. Jahn TR, Radford SE (2005) The Yin and Yang of protein folding. FEBS J 272(23):5962–5970. doi:10.1111/j.1742-4658.2005.05021.x
55. Fawzi NL, Chubukov V, Clark LA, Brown S, Head-Gordon T (2005) Influence of denatured and intermediate states of folding on protein aggregation. Protein Sci 14(4):993–1003. doi:10.1110/ps.041177505
56. Privalov PL (1996) Intermediate states in protein folding. J Mol Biol 258(5):707–725. doi:10.1006/jmbi.1996.0280
57. Jimenez JL, Guijarro JI, Orlova E, Zurdo J, Dobson CM, Sunde M, Saibil HR (1999) Cryo-electron microscopy structure of an SH3 amyloid fibril and model of the molecular packing. EMBO J 18(4):815–821. doi:10.1093/emboj/18.4.815
58. Makin OS, Serpell LC (2002) Examining the structure of the mature amyloid fibril. Biochem Soc Trans 30(4):521–525. doi:10.1042/
59. Serpell LC, Sunde M, Fraser PE, Luther PK, Morris EP, Sangren O, Lundgren E, Blake CC (1995) Examination of the structure of the transthyretin amyloid fibril by image reconstruction from electron micrographs. J Mol Biol 254(2):113–118. doi:10.1006/jmbi.1995.0604
60. Shivaprasad S, Wetzel R (2006) Analysis of amyloid fibril structure by scanning cysteine mutagenesis. Methods Enzymol 413:182–198. doi:10.1016/S0076-6879(06)13010-4
61. Shivji AP, Brown F, Davies MC, Jennings KH, Roberts CJ, Tendler S, Wilkinson MJ, Williams PM (1995) Scanning tunnelling microscopy studies of beta-amyloid fibril structure and assembly. FEBS Lett 371(1):25–28
62. Tycko R (2000) Solid-state NMR as a probe of amyloid fibril structure. Curr Opin Chem Biol 4(5):500–506
63. Bedrood S, Li Y, Isas JM, Hegde BG, Baxa U, Haworth IS, Langen R (2012) Fibril structure of human islet amyloid polypeptide. J Biol Chem 287(8):5235–5241. doi:10.1074/jbc.M111.327817
64. Fitzpatrick AW, Debelouchina GT, Bayro MJ, Clare DK, Caporini MA, Bajaj VS, Jaroniec CP, Wang L, Ladizhansky V, Muller SA, MacPhee CE, Waudby CA, Mott HR, De Simone A, Knowles TP, Saibil HR, Vendruscolo M, Orlova EV, Griffin RG, Dobson CM (2013) Atomic structure and hierarchical assembly of a cross-beta amyloid fibril. Proc Natl Acad Sci U S A 110(14):5468–5473. doi:10.1073/pnas.1219476110
65. Nelson R, Sawaya MR, Balbirnie M, Madsen AO, Riekel C, Grothe R, Eisenberg D (2005) Structure of the cross-beta spine of amyloid-like fibrils. Nature 435(7043):773–778. doi:10.1038/nature03680

66. Sachse C, Xu C, Wieligmann K, Diekmann S, Grigorieff N, Fandrich M (2006) Quaternary structure of a mature amyloid fibril from Alzheimer's Abeta(1-40) peptide. J Mol Biol 362(2):347–354. doi:10.1016/j.jmb.2006.07.011
67. Bonar L, Cohen AS, Skinner MM (1969) Characterization of the amyloid fibril as a cross-beta protein. Proc Soc Exp Biol Med 131(4):1373–1375
68. Kirschner DA, Abraham C, Selkoe DJ (1986) X-ray diffraction from intraneuronal paired helical filaments and extraneuronal amyloid fibers in Alzheimer disease indicates cross-beta conformation. Proc Natl Acad Sci U S A 83(2):503–507
69. Serpell LC, Berriman J, Jakes R, Goedert M, Crowther RA (2000) Fiber diffraction of synthetic alpha-synuclein filaments shows amyloid-like cross-beta conformation. Proc Natl Acad Sci U S A 97(9):4897–4902
70. Jahn TR, Makin OS, Morris KL, Marshall KE, Tian P, Sikorski P, Serpell LC (2010) The common architecture of cross-beta amyloid. J Mol Biol 395(4):717–727. doi:10.1016/j.jmb.2009.09.039
71. Balbirnie M, Grothe R, Eisenberg DS (2001) An amyloid-forming peptide from the yeast prion Sup35 reveals a dehydrated beta-sheet structure for amyloid. Proc Natl Acad Sci U S A 98(5):2375–2380. doi:10.1073/pnas.041617698
72. Diaz-Avalos R, Long C, Fontano E, Balbirnie M, Grothe R, Eisenberg D, Caspar DL (2003) Cross-beta order and diversity in nanocrystals of an amyloid-forming peptide. J Mol Biol 330(5):1165–1175
73. Sawaya MR, Sambashivan S, Nelson R, Ivanova MI, Sievers SA, Apostol MI, Thompson MJ, Balbirnie M, Wiltzius JJ, McFarlane HT, Madsen AO, Riekel C, Eisenberg D (2007) Atomic structures of amyloid cross-beta spines reveal varied steric zippers. Nature 447(7143):453–457. doi:10.1038/nature05695
74. Eisenberg D, Jucker M (2012) The amyloid state of proteins in human diseases. Cell 148(6):1188–1203. doi:10.1016/j.cell.2012.02.022
75. Goldschmidt L, Teng PK, Riek R, Eisenberg D (2010) Identifying the amylome, proteins capable of forming amyloid-like fibrils. Proc Natl Acad Sci U S A 107(8):3487–3492. doi:10.1073/pnas.0915166107
76. Lesne SE (2013) Breaking the code of amyloid-beta oligomers. Int J Cell Biol 2013:950783. doi:10.1155/2013/950783
77. Esteras-Chopo A, Serrano L, Lopez de la Paz M (2005) The amyloid stretch hypothesis: recruiting proteins toward the dark side. Proc Natl Acad Sci U S A 102(46):16672–16677. doi:10.1073/pnas.0505905102
78. Pastor MT, Esteras-Chopo A, Serrano L (2007) Hacking the code of amyloid formation: the amyloid stretch hypothesis. Prion 1(1):9–14
79. Hammarstrom P, Jiang X, Hurshman AR, Powers ET, Kelly JW (2002) Sequence-dependent denaturation energetics: a major determinant in amyloid disease diversity. Proc Natl Acad Sci U S A 99(Suppl 4):16427–16432. doi:10.1073/pnas.202495199
80. Hurshman Babbes AR, Powers ET, Kelly JW (2008) Quantification of the thermodynamically linked quaternary and tertiary structural stabilities of transthyretin and its disease-associated variants: the relationship between stability and amyloidosis. Biochemistry 47(26):6969–6984. doi:10.1021/bi800636q
81. Richardson JS, Richardson DC (2002) Natural beta-sheet proteins use negative design to avoid edge-to-edge aggregation. Proc Natl Acad Sci U S A 99(5):2754–2759. doi:10.1073/pnas.052706099
82. Solomon A (1986) Light chains of immunoglobulins: structural-genetic correlates. Blood 68(3):603–610
83. Carlsson M (1994) [150 years of Bence Jones protein. A reliable marker for multiple myeloma]. Lakartidningen 91(44):3993–3995
84. Glenner GG, Harbaugh J, Ohma JI, Harada M, Cuatrecasas P (1970) An amyloid protein: the amino-terminal variable fragment of an immunoglobulin light chain. Biochem Biophys Res Commun 41(5):1287–1289

85. Palladini G, Comenzo RL (2012) The challenge of systemic immunoglobulin light-chain amyloidosis (Al). Subcell Biochem 65:609–642. doi:10.1007/978-94-007-5416-4_22
86. Bellotti V, Mangione P, Merlini G (2000) Review: immunoglobulin light chain amyloidosis—the archetype of structural and pathogenic variability. J Struct Biol 130(2–3):280–289. doi:10.1006/jsbi.2000.4248
87. Ramirez-Alvarado M (2012) Amyloid formation in light chain amyloidosis. Curr Top Med Chem 12(22):2523–2533
88. Buxbaum J (1992) Mechanisms of disease: monoclonal immunoglobulin deposition. Amyloidosis, light chain deposition disease, and light and heavy chain deposition disease. Hematol Oncol Clin North Am 6(2):323–346
89. Rocken C, Hegenbarth V, Schmitz M, Stix B, Schade G, Mohnert A, Roessner A (2000) Plasmacytoma of the tonsil with AL amyloidosis: evidence of post-fibrillogenic proteolysis of the fibril protein. Virchows Arch 436(4):336–344
90. Solomon A, Weiss DT, Murphy CL, Hrncic R, Wall JS, Schell M (1998) Light chain-associated amyloid deposits comprised of a novel kappa constant domain. Proc Natl Acad Sci U S A 95(16):9547–9551
91. Yamamoto K, Yagi H, Lee YH, Kardos J, Hagihara Y, Naiki H, Goto Y (2010) The amyloid fibrils of the constant domain of immunoglobulin light chain. FEBS Lett 584(15):3348–3353. doi:10.1016/j.febslet.2010.06.019
92. Solomon A, Weiss DT, Williams TK (1992) Experimental model of human light-chain-associated disease. Curr Top Microbiol Immunol 182:261–267
93. Wall J, Murphy CL, Solomon A (1999) In vitro immunoglobulin light chain fibrillogenesis. Methods Enzymol 309:204–217
94. Klimtchuk ES, Gursky O, Patel RS, Laporte KL, Connors LH, Skinner M, Seldin DC (2010) The critical role of the constant region in thermal stability and aggregation of amyloidogenic immunoglobulin light chain. Biochemistry 49(45):9848–9857. doi:10.1021/bi101351c
95. Stenstad T, Magnus JH, Kolset SO, Cornwell GG 3rd, Husby G (1991) Macromolecular properties of glycosaminoglycans in primary AL amyloid fibril extracts of lymphoid tissue origin. Scand J Immunol 34(5):611–617
96. Stevens FJ, Kisilevsky R (2000) Immunoglobulin light chains, glycosaminoglycans, and amyloid. Cell Mol Life Sci 57(3):441–449
97. Jiang X, Myatt E, Lykos P, Stevens FJ (1997) Interaction between glycosaminoglycans and immunoglobulin light chains. Biochemistry 36(43):13187–13194. doi:10.1021/bi970408h
98. Martin DJ, Ramirez-Alvarado M (2011) Glycosaminoglycans promote fibril formation by amyloidogenic immunoglobulin light chains through a transient interaction. Biophys Chem 158(1):81–89. doi:10.1016/j.bpc.2011.05.011
99. McLaughlin RW, De Stigter JK, Sikkink LA, Baden EM, Ramirez-Alvarado M (2006) The effects of sodium sulfate, glycosaminoglycans, and Congo red on the structure, stability, and amyloid formation of an immunoglobulin light-chain protein. Protein Sci 15(7):1710–1722. doi:10.1110/ps.051997606
100. Ren R, Hong Z, Gong H, Laporte K, Skinner M, Seldin DC, Costello CE, Connors LH, Trinkaus-Randall V (2010) Role of glycosaminoglycan sulfation in the formation of immunoglobulin light chain amyloid oligomers and fibrils. J Biol Chem 285(48):37672–37682. doi:10.1074/jbc.M110.149575
101. Blancas-Mejia LM, Hammernik J, Marin-Argany M, Ramirez-Alvarado M (2014) Differential effects on light chain amyloid formation depend on mutations and type of glycosaminoglycans. J Biol Chem 290(8):4953–4965. doi:10.1074/jbc.M114.615401
102. Booth DR, Sunde M, Bellotti V, Robinson CV, Hutchinson WL, Fraser PE, Hawkins PN, Dobson CM, Radford SE, Blake CC, Pepys MB (1997) Instability, unfolding and aggregation of human lysozyme variants underlying amyloid fibrillogenesis. Nature 385(6619):787–793. doi:10.1038/385787a0

103. Sekijima Y, Hammarstrom P, Matsumura M, Shimizu Y, Iwata M, Tokuda T, Ikeda S, Kelly JW (2003) Energetic characteristics of the new transthyretin variant A25T may explain its atypical central nervous system pathology. Lab Invest 83(3):409–417
104. Sukhanova A, Poly S, Shemetov A, Bronstein I, Nabiev I (2012) Implications of protein structure instability: from physiological to pathological secondary structure. Biopolymers 97(8):577–588. doi:10.1002/bip.22055
105. Wei L, Berman Y, Castano EM, Cadene M, Beavis RC, Devi L, Levy E (1998) Instability of the amyloidogenic cystatin C variant of hereditary cerebral hemorrhage with amyloidosis, Icelandic type. J Biol Chem 273(19):11806–11814
106. Baden EM, Owen BA, Peterson FC, Volkman BF, Ramirez-Alvarado M, Thompson JR (2008) Altered dimer interface decreases stability in an amyloidogenic protein. J Biol Chem 283(23):15853–15860. doi:10.1074/jbc.M705347200
107. Poshusta TL, Katoh N, Gertz MA, Dispenzieri A, Ramirez-Alvarado M (2013) Thermal stability threshold for amyloid formation in light chain amyloidosis. Int J Mol Sci 14(11):22604–22617. doi:10.3390/ijms141122604
108. Raffen R, Dieckman LJ, Szpunar M, Wunschl C, Pokkuluri PR, Dave P, Wilkins Stevens P, Cai X, Schiffer M, Stevens FJ (1999) Physicochemical consequences of amino acid variations that contribute to fibril formation by immunoglobulin light chains. Protein Sci 8(3):509–517. doi:10.1110/ps.8.3.509
109. Wall J, Schell M, Murphy C, Hrncic R, Stevens FJ, Solomon A (1999) Thermodynamic instability of human lambda 6 light chains: correlation with fibrillogenicity. Biochemistry 38(42):14101–14108
110. Wetzel R (1997) Domain stability in immunoglobulin light chain deposition disorders. Adv Protein Chem 50:183–242
111. French DL, Laskov R, Scharff MD (1989) The role of somatic hypermutation in the generation of antibody diversity. Science 244(4909):1152–1157
112. Perfetti V, Ubbiali P, Vignarelli MC, Diegoli M, Fasani R, Stoppini M, Lisa A, Mangione P, Obici L, Arbustini E, Merlini G (1998) Evidence that amyloidogenic light chains undergo antigen-driven selection. Blood 91(8):2948–2954
113. Baden EM, Randles EG, Aboagye AK, Thompson JR, Ramirez-Alvarado M (2008) Structural insights into the role of mutations in amyloidogenesis. J Biol Chem 283(45):30950–30956. doi:10.1074/jbc.M804822200
114. Hurle MR, Helms LR, Li L, Chan W, Wetzel R (1994) A role for destabilizing amino acid replacements in light-chain amyloidosis. Proc Natl Acad Sci U S A 91(12):5446–5450
115. Dealwis C, Wall J (2004) Towards understanding the structure-function relationship of human amyloid disease. Curr Drug Targets 5(2):159–171
116. Gonzalez-Andrade M, Becerril-Lujan B, Sanchez-Lopez R, Cecena-Alvarez H, Perez-Carreon JI, Ortiz E, Fernandez-Velasco DA, del Pozo-Yauner L (2013) Mutational and genetic determinants of lambda6 light chain amyloidogenesis. FEBS J 280(23):6173–6183. doi:10.1111/febs.12538
117. Randles EG, Thompson JR, Martin DJ, Ramirez-Alvarado M (2009) Structural alterations within native amyloidogenic immunoglobulin light chains. J Mol Biol 389(1):199–210. doi:10.1016/j.jmb.2009.04.010
118. Souillac PO, Uversky VN, Millett IS, Khurana R, Doniach S, Fink AL (2002) Effect of association state and conformational stability on the kinetics of immunoglobulin light chain amyloid fibril formation at physiological pH. J Biol Chem 277(15):12657–12665. doi:10.1074/jbc.M109230200
119. Takahashi N, Hasegawa K, Yamaguchi I, Okada H, Ueda T, Gejyo F, Naiki H (2002) Establishment of a first-order kinetic model of light chain-associated amyloid fibril extension in vitro. Biochim Biophys Acta 1601(1):110–120
120. Wang F, Sen S, Zhang Y, Ahmad I, Zhu X, Wilson IA, Smider VV, Magliery TJ, Schultz PG (2013) Somatic hypermutation maintains antibody thermodynamic stability during affinity maturation. Proc Natl Acad Sci U S A 110(11):4261–4266. doi:10.1073/pnas.1301810110

121. Chiti F, Stefani M, Taddei N, Ramponi G, Dobson CM (2003) Rationalization of the effects of mutations on peptide and protein aggregation rates. Nature 424(6950):805–808. doi:10.1038/nature01891
122. Retter I, Althaus HH, Munch R, Muller W (2005) VBASE2, an integrative V gene database. Nucleic Acids Res 33(Database issue):D671–D674. doi:10.1093/nar/gki088
123. Ignatovich O, Tomlinson IM, Jones PT, Winter G (1997) The creation of diversity in the human immunoglobulin V(lambda) repertoire. J Mol Biol 268(1):69–77. doi:10.1006/jmbi.1997.0956
124. Collins AM, Wang Y, Singh V, Yu P, Jackson KJ, Sewell WA (2008) The reported germline repertoire of human immunoglobulin kappa chain genes is relatively complete and accurate. Immunogenetics 60(11):669–676. doi:10.1007/s00251-008-0325-z
125. Perfetti V, Casarini S, Palladini G, Vignarelli MC, Klersy C, Diegoli M, Ascari E, Merlini G (2002) Analysis of V(lambda)-J(lambda) expression in plasma cells from primary (AL) amyloidosis and normal bone marrow identifies 3r (lambdaIII) as a new amyloid-associated germline gene segment. Blood 100(3):948–953. doi:10.1182/blood-2002-01-0114
126. Bodi K, Prokaeva T, Spencer B, Eberhard M, Connors LH, Seldin DC (2009) AL-Base: a visual platform analysis tool for the study of amyloidogenic immunoglobulin light chain sequences. Amyloid 16(1):1–8. doi:10.1080/13506120802676781
127. Ozaki S, Abe M, Wolfenbarger D, Weiss DT, Solomon A (1994) Preferential expression of human lambda-light-chain variable-region subgroups in multiple myeloma, AL amyloidosis, and Waldenstrom's macroglobulinemia. Clin Immunol Immunopathol 71(2):183–189
128. Abraham RS, Geyer SM, Price-Troska TL, Allmer C, Kyle RA, Gertz MA, Fonseca R (2003) Immunoglobulin light chain variable (V) region genes influence clinical presentation and outcome in light chain-associated amyloidosis (AL). Blood 101(10):3801–3808. doi:10.1182/blood-2002-09-2707
129. Comenzo RL, Wally J, Kica G, Murray J, Ericsson T, Skinner M, Zhang Y (1999) Clonal immunoglobulin light chain variable region germline gene use in AL amyloidosis: association with dominant amyloid-related organ involvement and survival after stem cell transplantation. Br J Haematol 106(3):744–751
130. Solomon A, Frangione B, Franklin EC (1982) Bence Jones proteins and light chains of immunoglobulins. Preferential association of the V lambda VI subgroup of human light chains with amyloidosis AL (lambda). J Clin Invest 70(2):453–460
131. del Pozo Yauner L, Ortiz E, Sanchez R, Sanchez-Lopez R, Guereca L, Murphy CL, Allen A, Wall JS, Fernandez-Velasco DA, Solomon A, Becerril B (2008) Influence of the germline sequence on the thermodynamic stability and fibrillogenicity of human lambda 6 light chains. Proteins 72(2):684–692. doi:10.1002/prot.21934
132. Blancas-Mejia LM, Tellez LA, del Pozo-Yauner L, Becerril B, Sanchez-Ruiz JM, Fernandez-Velasco DA (2009) Thermodynamic and kinetic characterization of a germ line human lambda6 light-chain protein: the relation between unfolding and fibrillogenesis. J Mol Biol 386(4):1153–1166. doi:10.1016/j.jmb.2008.12.069
133. Khurana R, Gillespie JR, Talapatra A, Minert LJ, Ionescu-Zanetti C, Millett I, Fink AL (2001) Partially folded intermediates as critical precursors of light chain amyloid fibrils and amorphous aggregates. Biochemistry 40(12):3525–3535
134. Solomon A, Weiss DT, Kattine AA (1991) Nephrotoxic potential of Bence Jones proteins. N Engl J Med 324(26):1845–1851. doi:10.1056/NEJM199106273242603
135. Solomon A, Weiss DT, Pepys MB (1992) Induction in mice of human light-chain-associated amyloidosis. Am J Pathol 140(3):629–637
136. Sikkink LA, Ramirez-Alvarado M (2008) Biochemical and aggregation analysis of Bence Jones proteins from different light chain diseases. Amyloid 15(1):29–39. doi:10.1080/13506120701815324
137. Frare E, Polverino De Laureto P, Zurdo J, Dobson CM, Fontana A (2004) A highly amyloidogenic region of hen lysozyme. J Mol Biol 340(5):1153–1165. doi:10.1016/j.jmb.2004.05.056

138. Iconomidou VA, Leontis A, Hoenger A, Hamodrakas SJ (2013) Identification of a novel 'aggregation-prone'/'amyloidogenic determinant' peptide in the sequence of the highly amyloidogenic human calcitonin. FEBS Lett 587(6):569–574. doi:10.1016/j.febslet.2013.01.031
139. Ivanova MI, Thompson MJ, Eisenberg D (2006) A systematic screen of beta(2)-microglobulin and insulin for amyloid-like segments. Proc Natl Acad Sci U S A 103(11):4079–4082. doi:10.1073/pnas.0511298103
140. Tenidis K, Waldner M, Bernhagen J, Fischle W, Bergmann M, Weber M, Merkle ML, Voelter W, Brunner H, Kapurniotu A (2000) Identification of a penta- and hexapeptide of islet amyloid polypeptide (IAPP) with amyloidogenic and cytotoxic properties. J Mol Biol 295(4):1055–1071. doi:10.1006/jmbi.1999.3422
141. Thompson MJ, Sievers SA, Karanicolas J, Ivanova MI, Baker D, Eisenberg D (2006) The 3D profile method for identifying fibril-forming segments of proteins. Proc Natl Acad Sci U S A 103(11):4074–4078. doi:10.1073/pnas.0511295103
142. Tsolis AC, Papandreou NC, Iconomidou VA, Hamodrakas SJ (2013) A consensus method for the prediction of 'aggregation-prone' peptides in globular proteins. PLoS One 8(1):e54175. doi:10.1371/journal.pone.0054175
143. Kabat EA, Wu TT, Perry HM, Gottesman KS, Foeller C (1991) Sequences of proteins of immunological interest, vol 5, 5th edn. National Institutes of Health, Bethesda
144. Beerten J, Van Durme J, Gallardo R, Capriotti E, Serpell L, Rousseau F, Schymkowitz J (2015) WALTZ-DB: a benchmark database of amyloidogenic hexapeptides. Bioinformatics 31(10):1698–1700. doi:10.1093/bioinformatics/btv027
145. Maurer-Stroh S, Debulpaep M, Kuemmerer N, Lopez de la Paz M, Martins IC, Reumers J, Morris KL, Copland A, Serpell L, Serrano L, Schymkowitz JW, Rousseau F (2010) Exploring the sequence determinants of amyloid structure using position-specific scoring matrices. Nat Methods 7(3):237–242. doi:10.1038/nmeth.1432
146. del Pozo Yauner L, Ortiz E, Becerril B (2006) The CDR1 of the human lambdaVI light chains adopts a new canonical structure. Proteins 62(1):122–129. doi:10.1002/prot.20779

Thermodynamics and Kinetics of Amyloid Aggregation from Atomistic Simulations

Bogdan Barz and Birgit Strodel

Abstract A major cause for cellular toxicity involved in the onset of several neurodegenerative diseases is the aberrant aggregation of peptides or proteins into oligomers and eventually fibrils. In the case of Alzheimer's disease, the main aggregating peptide is the amyloid β-peptide with two main alloforms of 40 (Aβ40) and 42 (Aβ42) amino acids. Numerous experimental studies have shown that early oligomers on-pathway to fibril formation are toxic, with Aβ42 showing a higher toxicity than Aβ40. To explore the aggregation mechanisms and differences in the oligomeric conformations we follow the aggregation of Aβ40 and Aβ42 from isolated monomers using all-atom molecular dynamics simulations. We describe the kinetics of aggregation and differences in the pathways arising from sequence differences using transition networks.

This chapter is partly adapted from Barz et al. (J Phys Chem B 118(4):1003, 2014; Chem Commun 50:5373, 2014) with permission from the American Chemical Society and The Royal Society of Chemistry, respectively.

Keywords Amyloid peptides • Protein aggregation • Molecular dynamics • All-atom simulations • Free energy • Transition network • Disconnectivity graph • First passage time • Alzheimer's disease • Amyloid-beta(1-42) • Oligomers

B. Barz (✉) • B. Strodel
Forschungszentrum Jülich, Institute of Complex Systems: Structural Biochemistry (ICS-6), 52425, Jülich, Germany
e-mail: b.barz@fz–juelich.de; b.strodel@fz-juelich.de

L. Olivares-Quiroz et al. (eds.), *Physical Biology of Proteins and Peptides*,
DOI 10.1007/978-3-319-21687-4_2

2.1 Transition Networks for Modeling the Kinetics of Protein Aggregation

2.1.1 Introduction

Amyloid protein aggregation is one of the main factors involved in the onset of several neurodegenerative conditions, including Alzheimer's, Parkinson's, and prion diseases. The aberrant assembly of proteins such as amyloid-β (Aβ), α-synuclein, and the prion protein into fibrils has been associated with the death of neuronal cells. There is increasing evidence that oligomers, present in the early stages of aggregation, are the neurotoxic species, rather than highly organized fibrils [1–4]. The prion protein and Aβ are two examples of amyloidogenic proteins whose early assembly mechanisms have been investigated using both experimental [3, 5, 6] and computational tools [7–10]. Two short fragments from these proteins exhibit aggregation behavior and cross-β structures similar to the ones specific to fibrils formed by the full-length proteins. One is the GNNQQNY fragment from the N-terminal prion determining region of the yeast protein Sup35 (abbreviated GNNQQNY). The second is the KLVFFAE fragment from Aβ (abbreviated KLVFFAE). The motivation for studying these shorter fragments of amyloidogenic proteins is the hypothesis that their behavior captures essential features of the original proteins [11]. Moreover, these two peptides have very different primary structures: GNNQQNY is polar, while KLVFFAE has a hydrophobic core. They are therefore likely to exhibit different aggregation pathways, whose elucidation would contribute to a better understanding of the sequence–aggregation relationship.

One of the methods for describing stochastic biological processes involves kinetic transition networks [12, 13] and disconnectivity graphs [14, 15], which has been applied to the study of conformational dynamics of peptides or protein folding [16–19] and recently to aggregation [20–22]. In our recent work [20] we characterized the assembly of six monomers into one hexamer for GNNQQNY and KLVFFAE, based on molecular dynamics (MD) simulations, and compared the pathways. To provide a detailed picture of the aggregation process, we introduced aggregation states defined on the basis of certain structural characteristics, in addition to oligomer size, and derived kinetic transition networks (KTNs) illustrating the population and transitions between these states [20]. As these KTNs are rather complex, we applied the max-flow min-cut procedure, which removes less important nodes and edges of the KTNs while preserving the maximum flow [16, 23–25]. The resulting networks, which we denote mincut transition networks (MTNs), provide a concise picture of the aggregation pathway in terms of stable aggregation states and transitions between them. However, they lack kinetic information, which we included by describing the free energy landscape for aggregation through disconnectivity graphs (DGs) [14, 15] using the node and edge data from the MTNs. Although both peptides GNNQQNY and KLVFFAE have been previously studied with both coarse-grained [26–29] and all-atom models [13, 22, 30–32], our study [20] was the first comparison of their aggregation behavior using a kinetic description that involves transition networks.

2.1.2 Materials and Methods

MD Simulations For both GNNQQNY and KLVFFAE we performed five independent all-atom MD simulations of six peptides inserted in a cubic box with side length 10 nm filled with explicit SPC water [33], resulting in a concentration of about 10 mM. Each simulation was 500 ns long, producing a total of 2.5 μs per sequence. The simulations were performed with the Gromacs 4.5.5 parallel software package [34] and the GROMOS96 43A1 force field [35]. Hydrogen atoms were treated as virtual interaction sites, permitting an integration time step of 5 fs while maintaining energy conservation [36]. The system was maintained at 310 K and 1 bar via velocity rescaling with a stochastic term algorithm [37] and the Berendsen barostat [38], respectively. Electrostatic interactions were described using the particle mesh Ewald algorithm [39, 40] with a cutoff of 0.9 nm. Van der Waals interactions were cutoff at 1.4 nm with a switching distance of 0.9 nm. During each MD simulation, structures were saved every 5 ps, resulting in 500,000 snapshots per peptide.

Definition of Aggregation States The aggregation state is defined as a number with five digits, $N1N2N3N4N5$, where each digit corresponds to a structural feature or to an oligomeric state. $N1$ represents the oligomeric size, identified using a cutoff distance of 0.5 nm between any two atoms belonging to different peptides. $N2$ is the number of peptides connected by at least one hydrogen-bond. Hydrogen-bonds are defined based on distance and angle cutoffs of 0.41 nm and 40°, respectively. $N3$ and $N4$ are the number of antiparallel and parallel peptides within the oligomer. Two chains are antiparallel or parallel if they are connected by at least two hydrogen-bonds involving two unique amino acids from each peptide. $N5$ is the average number of amino acids in β-strand conformation per peptide in the oligomer. An amino acid is defined to be in a β-strand conformation if the dihedral angles ϕ and ψ (in degree) of the backbone are contained in the polygon with vertices (−180, 180), (−180, 126), (−162, 126), (−162, 108), (−144, 108), (−144, 90), (−50, 90), (−50, 180) [41]. A definition purely based on backbone torsional angles allows to assign β-strand conformation to peptide monomers in an extended conformation, which would not be possible with algorithms such as STRIDE [42] and DSSP [43] that are based on the hydrogen bond energy for the assignment of protein secondary structure. Furthermore, stricter definitions based on ϕ and ψ have been tested. For instance, in [44] for a β-strand it was assumed that $-150^\circ \leq \phi \leq -90^\circ$ and $90^\circ \leq \psi \leq 150^\circ$. While the amount of β-strands identified with this stricter definition is smaller than with the definition applied in the current work, both definitions provide qualitatively similar results. One can thus assume that our state determination would not be affected by a stricter definition of $N5$.

Transition Matrix To calculate the transition matrix that includes all pairwise transitions between aggregation states we first identified all the states and the number of transitions between them along the 2.5 μs trajectory using a lag time of 5 ps. Using these states and transitions we built an $N \times N$ matrix, where N is the number of states encountered. Each element of the matrix represents the population of a particular transition between two states. After normalizing the rows, each element represents the probability for a particular transition to occur. The transition matrix did not change significantly for longer lag times.

Kinetic Transition Networks The row normalized transition matrices were converted into KTNs using the software Visone [45] and the minimum stress algorithm. In the KTN plots, the nodes represent aggregation states, the area of each node is proportional to the population of the state, and the color of the node indicates the average number of amino acids in the β-strand conformation ($N5$). The color of network edges corresponds to the value of the transition probability. Based on KTNs we have derived networks that preserve the maximum flow using the mincut procedure, which we denote as MTNs [16, 23–25]. In the case of MTNs the thickness of the edges corresponds to the number of transitions and the color of the nodes to the quantity ($N3 - N4$).

Representative States and Conformations For each KTN we selected representative aggregation states and illustrate representative conformations in Fig. 1. To select representative states we first sorted by decreasing population and then identified the states with the highest population for each $N1$ value (or oligomer). For each state we selected a structure that reflects characteristics of the aggregation state. The snapshots were generated with the software package VMD [46].

Disconnectivity Graphs The number of times state i was visited during the simulations was used to define the partition function as $Z_i = \sum_j n_{ij}$, where n_{ij} is the number of transitions from state j to state i [16, 24]. The free energy of state i was derived from Z_i as $F_i = -kT \ln Z_i$, where k is the Boltzmann constant and T is the temperature. To calculate the free energy barrier between two states we performed the following steps: (1) we derived the undirected graph with edge capacities $c_{ij} \sim (n_{ij} + n_{ji})/2$, (2) we applied the mincut algorithm to the new undirected network to obtain the new transition capacities $c'_{ij} = c'_{ji}$, which reflect the maximum flow through the network, (3) we derived the partition function $Z_{ij} = c'_{ij}(h/kT)(1/t_s)$, where h is the Planck constant, T is 310 K and t_s is the sampling time interval of 5 ps, (4) we estimated the free energy barriers as $F_{ij} \sim -kT \ln(Z_{ij})$ [24]. Using the minima and barriers we constructed free energy disconnectivity graphs based on the method developed by Becker and Karplus [14].

2.1.3 Results and Discussion

In the case of GNNQQNY, hexamers are the dominant oligomers at 310 K, with a population of 77.2 %, followed by monomers (11.9 %), pentamers (4.8 %), dimers (3 %), trimers (2.3 %), and tetramers (0.7 %). For KLVFFAE, the hexamers are also the most populated oligomeric state (65.4 %), followed by monomers (15.2 %), dimers (11.0 %), tetramers (3.9 %), pentamers (2.9 %), and trimers (1.6 %). The high hexamer population indicates fast initial assembly of small oligomers into a hexamer that undergoes slower rearrangements. The fact that GNNQQNY has a larger population of hexamers (77.2 %) than KLVFFAE (65.4 %) indicates faster assembly, which might result from the presence of a tyrosine residue [30]. To obtain a better picture of the assembly process we now consider kinetic transition networks.

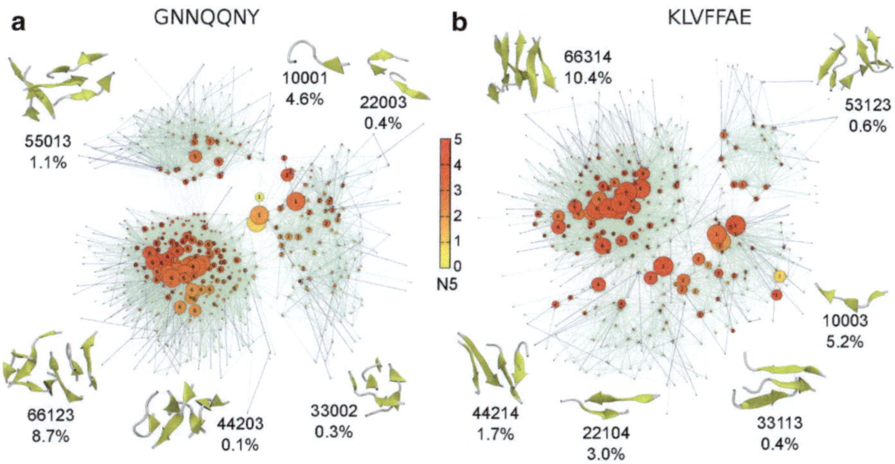

Fig. 1 Kinetic transition networks for GNNQQNY (**a**) and KLVFFAE (**b**). For both peptides the aggregation states *N*1*N*2*N*3*N*4*N*5 are shown as nodes connected by edges that represent transition probabilities. The size of each node corresponds to the population and the *color* corresponds to the average number of β-strand amino acids, *N*5. The *color* of the edges reflects the value of the transition probability, from *light green* for small probabilities to *dark blue* for high probabilities. For each oligomer size, *N*1, we illustrate representative conformations corresponding to the aggregation state with the highest population, together with the population of that state in percentage. Reprinted with permission from [20]. Copyright 2014 American Chemical Society

Kinetic Transition Networks KTNs for GNNQQNY and KLVFFAE, based on the aggregation state definition (*N*1*N*2*N*3*N*4*N*5) from the "Materials and Methods" section, are shown in Fig. 1a, b, respectively. In both cases there are three distinct interconnected clusters: C1 for states with *N*1 (oligomer size) values of 1–4, C2 for states with *N*1 = 5, and C3 for *N*1 = 6. For KLVFFAE the clusters are more interconnected than for GNNQQNY, which suggests a more dynamic assembly process, with larger oligomers breaking into smaller ones more frequently. The coloring of the nodes is based on the average number of β-strand amino acids per peptide, encoded in *N*5. This information reveals an overall higher β-strand content in KLVFFAE relative to GNNQQNY, even for the states with low *N*1 values, indicating a higher degree of organization during the early assembly stages for KLVFFAE. It is important to note the central location of the states with *N*1 = 1 (e.g. 10,001) in the case of GNNQQNY, but not for KLVFFAE. This result agrees with the CGTNs, where we observed that monomers contribute to all oligomeric states for GNNQQNY but not for KLVFFAE.

While both C1 clusters include states with different *N*1 values, the structure for GNNQQNY is more compact and has fewer transitions to cluster C3 than for KLVFFAE. Representative aggregation states for GNNQQNY are 10,001 with a population of 4.6 %, 22,003 (0.4 %), 33,002 (0.3 %), and 44,203 (0.1 %). Representative conformations for each state are shown in Fig. 1a. They generally have a small number of parallel or antiparallel peptide pairs and an average number of β-strand amino acids below three. In the case of KLVFFAE representative aggregation states from cluster C1 are 10,003 (2.5 %), 22,104 (3.0 %), 33,113 (0.4 %), and 44,214 (1.7 %). For most of these states the number of parallel and antiparallel peptide pairs is larger than for GNNQQNY and the number of antiparallel pairs is higher than the number of parallel peptide pairs. In addition, the higher average number of β-strand amino acids indicates more ordered small oligomers for KLVFFAE (Fig. 1b).

The C2 clusters contain only states with $N1=5$, with a larger population for GNNQQNY. The representative aggregation states for this cluster are 55,013 with a population of 1.1 % for GNNQQNY and 53,123 (0.6 %) for KLVFFAE. Both states have more parallel pairs than antiparallel, with more pairs in total for KLVFFAE (antiparallel:parallel = 1:2) than for GNNQQNY (0:1) and thus more β-sheet-like structures. However, GNNQQNY has more chains connected by hydrogen-bonds. The most populated cluster for both fragments is C3, containing states with $N1=6$. The most representative aggregation states for this cluster are 66,123 (8.7 %) for GNNQQNY and 66,314 (10.4 %) for KLVFFAE. GNNQQNY samples a higher average number of parallel pairs (1:2), while for KLVFFAE the states with more antiparallel pairs are more common (3:1). In terms of β-strand content, cluster C3 shows a gradual transition from low to high for GNNQQNY, while KLVFFAE has a more uniform distribution of higher β-strand content. State 66,314 of KLVFFAE has 4 β-strand amino acids on average.

Mincut Transition Networks The KTNs described in the previous section provide a detailed picture of the aggregation process. However, the large number of transitions and nodes make it difficult to pinpoint the main pathways that dominate the assembly process. To identify the maximum flow in the transition network we applied the minimum cut method (mincut) and derived MTNs.

The MTNs of the two peptides are shown in Fig. 2, where we identify the states with the same value of $N1$ as basin Bi, with $i=1,\ldots,6$. For GNNQQNY basin B6 has one central node corresponding as state 66,123, which is connected radially to the other states. The highly populated states from B6 have an edge of high capacity in direct contact with the central node, and some of them have one to three edges directly connected to less populated nodes. The GNNQQNY assemblies from basin B6 have a higher propensity to form parallel pairs, but also contain a significant number of antiparallel pairs. State 66,123 is also the one that connects B6 to the other basins. An important observation is that the transition to/from B6 from/to B2, B3 and B4 occurs through B1, specifically through state 10,002, which is not the highest populated state within B1. This observation indicates the central role of monomers in the assembly process for GNNQQNY.

The MTN of KLVFFAE is shown in Fig. 2b. The topology of basin B6 is very different from that of GNNQQNY. Rather than having one central node surrounded radially by the other nodes, KLVFFAE displays several nodes that are at the center of local clusters. One can identify five such nodes connected to the central node, corresponding to state 66,314. Three of four nodes are among the top ten states with the highest population (Table 1). The difference in antiparallel to parallel pairs ($N3-N4$) shows a clear shift from negative to positive values for the five local clusters.

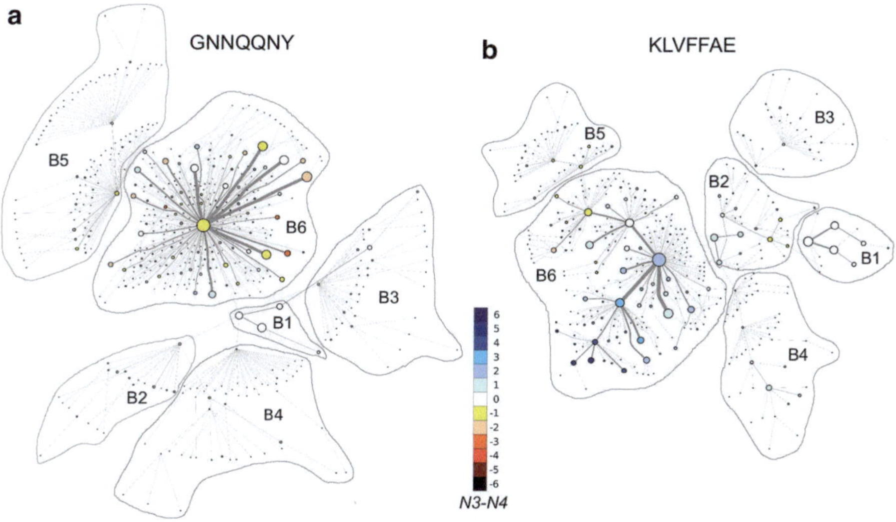

Fig. 2 Mincut transition networks for GNNQQNY (**a**) and KLVFFAE (**b**). The coloring of the nodes reflects *N*3 – *N*4, which indicates the excess number of antiparallel peptide pairs relative to parallel peptide pairs. Edges shown in *gray* represent the number of transitions returned by the mincut algorithm. Clusters of states with identical values of *N*1 have been outlined and identified as basin B*i*, where *i* takes the value of *N*1. Reprinted with permission from [20]. Copyright 2014 American Chemical Society

Table 1 The ten aggregation states with the highest populations for GNNQQNY, and KLVFFAE

GNNQQNY		KLVFFAE	
Aggregation state	Population (%)	Aggregation state	Population (%)
66,123	8.6	66,314	10.4
66,023	6.2	10,003	5.2
66,124	5.6	66,214	5.0
66,013	5.2	66,304	4.4
66,113	4.9	66,224	4.0
10,001	4.6	10,004	3.9
66,223	3.7	10,002	3.6
10,002	3.3	66,124	3.2
66,213	2.9	66,324	3.1
66,033	2.1	22,104	3.0

For each peptide the aggregation state *N*1*N*2*N*3*N*4*N*5 is shown in the left column with the percentage population in the right column. Reprinted with permission from [20]. Copyright 2014 American Chemical Society

This shift corresponds to a gradual transition from more parallel to more antiparallel pairs. Overall, the nodes in basin B6 have a higher propensity for antiparallel pairs. Interestingly, the nodes with more parallel pairs are connected to basin B5 through state 66,124, while the majority of nodes with more antiparallel pairs are closer and connected to basins B1–B4 through state 66,314. Basins B1, B3 and B4 are connected to basin B6 through B2 (state 22,103), suggesting that dimers play a central role in the pathway to hexamers.

One major difference between the above networks is the complex nature of basin B6, which suggests distinct paths for evolution of the collapsed hexamer in the two peptides. The distribution of parallel and antiparallel pairs is another characteristic of the two B6 basins. Numerous studies have reported a preference for parallel β-strands in GNNQQNY [26, 29–31] and antiparallel β-strands in KLVFFAE [32], in agreement with our findings. The central role of monomers in the assembly of small oligomers into hexamers is also clear from the MTN of GNNQQNY, in contrast to the importance of dimers for KLVFFAE. A two stage "dock-lock" mechanism [32] for monomer addition to preformed structured oligomers was previously proposed for KLVFFAE fibril formation. Our results suggest that pathways involving dimers might also be relevant.

To explore the role of electrostatic interactions during oligomer formation we identified salt bridges between amino acids K and E from different KLVFFAE peptides. We found one, two and three salt bridges present in the system for 23.6, 4.3 and 0.2 % of the simulation time, respectively. Given that we did not observe in-register antiparallel β-sheets, as found in the KLVFFAE fibrils [47], and that the salt bridges are not as abundant as in a computational study of pre-formed KLVFFAE fibrils [48], we suggest that salt bridges play a secondary role in the early aggregation of KLVFFAE, i.e. they are not as important as in fibril formation and stability.

While we did not observe actual fibrils stabilized by steric zippers in our simulations, we did find KLVFFAE hexamers resembling β-barrel structures (Fig. 1b, state 66,314). A similar hexamer arrangement has been recently observed in experiments performed by Laganowsky et al. and Liu et al. [49, 50] when studying the aggregation of segments of the amyloid-forming protein αB crystallin. In these studies they report cylindrical oligomers (cylindrins) formed by six out-of-register β-sheets as early oligomers off-pathway in terms of in-register fibril formation, but on-pathway for less stable out-of-register fibrils. The oligomers reported by Laganowsky et al. were found to be toxic, and the β-barrel conformation was suggested as a possible template for toxic amyloidogenic oligomers. The stability of the cylindrins was later studied computationally by Berhanu et al. [51]. A recent computational study confirmed the existence of β-barrel oligomers containing eight KLVFFAE peptides [52], similar to the hexamers observed in our study. In the case of KLVFFAE the out-of-register arrangement of β-sheets has been observed experimentally and shown to lead to stable nanotubular structures, rather than the classical amyloid fiber with cross-β structure [53]. The fact that we do not observe fibrils, in contrast to Berhanu et al. [48], is probably due to the limited number of peptides used in our study and also to the relatively short simulation time compared to the time necessary for fibril formation, which in vitro requires minutes or even hours. Moreover, the hexamers with β-barrel like structure observed in our study are indeed off-pathway for in-register fibrils, and there might be a rather high energy barrier to formation of the in-register β-sheet fibrils, as indicated by Laganowsky et al. [49].

Disconnectivity Graphs Disconnectivity graphs (DG) were originally introduced to describe the dynamics of a small peptide [14] and were subsequently employed to gain insight into the emergence of observable properties from the underlying energy landscape in a wide variety of systems [15]. A disconnectivity graph analysis focuses on local minima and the transition states that connect them [14]. It can also be understood as a clustering according to energetic barriers, where the energy landscape is partitioned into 'superbasins' [14]. Minima are in the same basin if they are connected by pathways with energy barriers below a particular energy threshold ΔE , which is defined by the user and determines the coarseness of the DG. Two minima are in different basins if the barrier along the pathway interconverting these two minima exceeds ΔE . Thus, the smaller ΔE the more superbasins appear in the DG, while more local minima are clustered into superbasins with increasing ΔE . Furthermore, for higher energies more superbasins appear as more local minima become accessible. DGs can be used to represent potential-energy and free-energy surfaces and are a useful diagnostic in visually assessing the thermodynamic and kinetic behavior of a system as they reproduce both the energies of minima and the barriers that separate them. The idea of DGs becomes clearest when comparing the disconnectivity graph for three different, archetypal energy landscapes [15] (Fig. 3).

In the present work we have constructed DGs for the aggregation of GNNQQNY and KLVFFAE based on the method described by Krivov et al. [24] (Fig. 4). In Fig. 4a one can identify several free energy funnel for GNNQQNY with the lowest value at 310 K for state 66123, followed in order of increasing free energy by 10,001, 55,013, 22,003, 33,002, and 44,203. For KLVFFAE the state with the lowest free energy at 310 K is 66,314, followed by states 10,003, 22,104, 44,214, 53,123 and 33,113. These orders correlate with the populations of the states given in Table 1, which is plausibe as the partition function Z_i is proportional to the population of state *i*. For GNNQQNY and KLVFFAE the hexamer state 66,123 and 66,314, respectively, is the most populated state and thus is the global free minimum in the corresponding DG. After the hexamer, a monomeric state is most stable for both GNNQQNY and KLVFFAE (10,001 for GNNQQNY and 10,003 for KLVFFAE) and thus shows up as second lowest energy funnel in both DGs. Here, a difference is that 10,003 is the second most populated state for KLVFFAE and thus corresponds to the second lowest energy minimum in the DG while for GNNQQNY 10,001 is only the sixth most populated state. Therefore, the hexameric energy funnel involves more local minima with free energies below that for the monomeric GNNQQNY state. Further differences appear to show for the other oligomer sizes. For instance, for KLVFFAE the dimer is rather stable (22,014 is the tenth most populated state) and thus corresponds to the third lowest energy funnel while for GNNQQNY the dimer is less and the pentamer more stable. Comparing the DGs for GNNQQNY and KLVFFAE with those for the archetypal energy landscapes in Fig. 3 we can conclude that the aggregation of KLVFFAE is best described by the 'palm tree' while the aggregation of GNNQQNY produces a DG with characteristics of both the 'palm tree' representing the individual oligomer sizes and the 'weeping willow' for the overall shape of the DG.

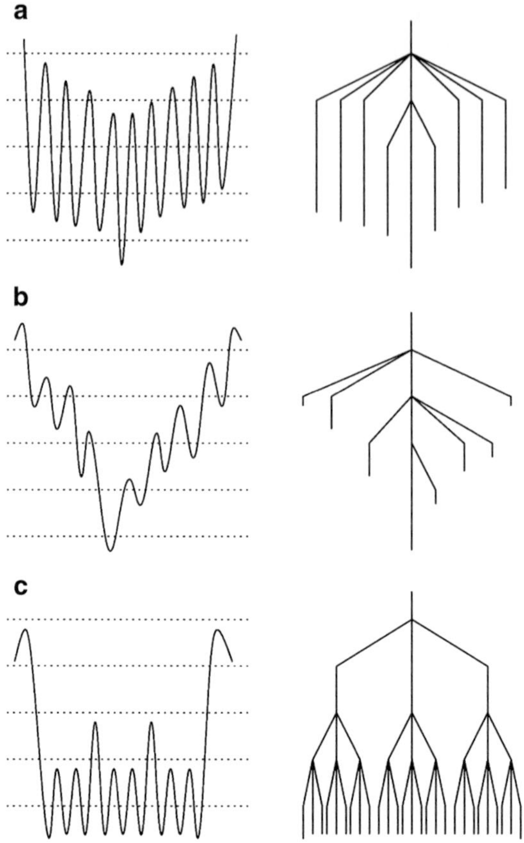

Fig. 3 Pictorial correspondence between the energy surface and the disconnectivity graph for three different energy landscapes [15]. Graphs, which are connected but contain no cycles, are known as 'trees'. Thus, different energy landscapes can produce different tree types. (**a**) The 'weeping willow' results from a gentle funnel with large barriers. (**b**) The 'palm tree'results from a steeper funnel with lower barriers. (**c**) The 'banyan tree' results from a rough landscape. Reprinted by permission from Macmillan Publishers Ltd: Nature [15], copyright (1998)

The DGs also allow to deduce information about the energy barriers involved in peptide aggregation. Figure 4c indicates a free energy barrier for the GNNQQNY trimer formation from monomers, or to state 33,002 from state 10,001, of about 5.4 kcal/mol. This value is slightly higher than the barriers (around 3 kcal/mol) between disordered aggregates and in-register mixed parallel/antiparallel aggregates reported by Gsponer et al. [30]. In the case of dimer formation, or transitions to state 22,003 from 10,001, the free energy barrier is around 5.1 kcal/mol. Barriers of 2 kcal/mol and 3 kcal/mol for GNNQQNY dimer formation have been reported previously by Reddy et al. [54] and Strodel et al. [13]. The higher barrier observed in our simulations might be due to the fact that the monomers are initially spatially separated in explicit solvent. Overall, GNNQQNY has higher barriers for transitions

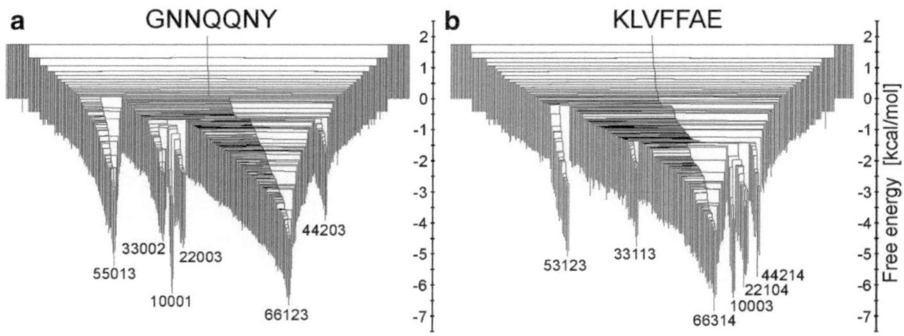

Fig. 4 Mincut disconnectivity graphs for GNNQQNY (**a**) and KLVFFAE (**b**). The vertical axis is $F_i \sim -kT \ln Z_i$ and $F_{ij} \sim -kT \ln Z_{ij}$ for minima and barriers, respectively. Representative aggregation states for the lowest free energy minima are labelled and corresponding structures are shown in Fig. 1. Reprinted with permission from [20]. Copyright 2014 American Chemical Society

to the most stable aggregation state than KLVFFAE in all cases, except for states with $N1 = 2$ and 4. This is an interesting result in the context of the earlier observation that monomers and dimers are central to the assembly process in GNNQQNY and KLVFFAE, respectively. In the case of GNNQQNY the higher free energy barrier for aggregation might be due to the polar nature of the peptide, while for KLVFFAE the fact that the barrier is higher for dimers and tetramers might be related to the reduced hydrophobic surface compared to monomers.

2.1.4 Conclusions

In this work [20] we highlighted differences between the aggregation pathways of two peptides with alternative amino acid sequences. While the two fragments might reflect the aggregation behavior of the full proteins, the relation between sequence and assembly mechanism is also of general interest. The early aggregation process of the two protein fragments from a large number of monomers has been studied before in silico, mostly using coarse-grained methods [26–29]. More detailed all-atom studies have focused on the assembly of two or three monomers [13, 22, 30, 32, 54, 55], the addition of monomers to a preformed oligomer [32, 56, 57], or the stability of preordered fibrilar structures [31, 58, 59]. In [20] we investigated the aggregation of six monomers using all-atom MD simulations in explicit solvent.

Detailed kinetic transition networks indicate that monomers are the most important species in the aggregation of GNNQQNY. A higher β-strand content for all KLVFFAE assemblies compared to GNNQQNY is observed, due to the hydrophobic residues, while in the case of the GNNQQNY hexamer a gradual transition from low to high β-strand content occurs. Mincut transition networks reveal a radial distribution of aggregation states of GNNQQNY for hexamer rearrangements dominated by parallel peptide pairs, and a modular distribution in the case of KLVFFAE, with a gradual increase in the relative number of antiparallel to parallel peptide pairs. Disconnectivity graphs reveal higher free energy barriers for hexamer formation from monomers, trimers and pentamers of GNNQQNY compared to KLVFFAE, and from dimers and tetramers of KLVFFAE compared to GNNQQNY. This result is probably due to the hydrophilic/hydrophobic nature of the two peptides and emphasizes the importance of the stability of monomers and dimers for the aggregation of GNNQQNY and KLVFFAE, respectively.

The observation that monomers play a key role in the aggregation of GNNQQNY is consistent with previous studies indicating that the main pathway for the prion protein Sup35 propagation is monomer addition [60]. Moreover, the reduced β-strand content in the intermediate oligomers of GNNQQNY observed in our study agrees with a previous observation that oligomers of Sup35 lack β-strand structure before being incorporated into fibrils [61]. For KLVFFAE the observation that dimers play an important role in aggregation process is consistent with the contribution of dimers and hexamers to aggregation of Aβ peptide (1–42) [6, 7]. Furthermore, the high β-strand content of the KLVFFAE oligomers in our study is consistent with a higher β-strand propensity in Aβ peptide oligomers [7]. In conclusion, our study [20] highlights key differences in the assembly process of two aggregation-prone peptides with contrasting amino acid sequences, as well as connections to the aggregation of the corresponding full-length proteins. In the next part it will be shown how this approach helps elucidating the aggregation process of full-length Aβ.

2.2 Modeling Oligomer Formation of Alzheimer's Aβ Peptide

2.2.1 Introduction

In the following it is described for the Alzheimer's Aβ peptide how the formation of amyloid oligomers can be assessed via molecular simulations. Oligomers of Aβ are considered one of the main causes of neurotoxicity and are thus highly associated with the onset of Alzheimer's disease [62]. Experimental methods are able to identify some characteristics of aggregating proteins such as the oligomer size distribution or cellular toxicity [6], but due to the fast conversion of oligomers into fibrils, the elucidation of their structure at molecular level is challenging. Computational methods have the advantage of atomistic detail but are exposed to the challenge that the size and time scales in experiment are usually larger than in atomistic simulations. In a recent study we reported the early assembly of Aβ42 at experimental concentrations using all-atom molecular dynamics (MD) simulations in implicit solvent, which were initiated from 20 isolated Aβ42 monomers [21]. To describe the assembly process we derived a maximum flow transition network (MTN) based on aggregation states defined by $N1|N2|N3$, where $N1$ represents the oligomeric size, $N2$ is the average number of hydrogen-bonds between individual chains from the oligomer, and $N3$ is the average number of amino acids in β-strand conformation per peptide in the oligomer. Detailed information about the methods are provided above and in [20, 21].

2.2.2 Results

Transition Network The MTN in Fig. 5 displays a complex aggregation process in which initial monomers assemble into oligomers up to 18-mers during 200 ns. The aggregation states with $N1 = 1$ are distributed linearly with a gradual increase in β-strand content from right to left. The monomers with more β-strands more readily aggregate than the others: state 1|0|17 is the main connection node to oligomers 2|2|15, 3|4|16, 4|6|13, 5|6|17, 12|7|13 and 17|8|13. These states are also the central connection to the other states with the same $N1$ value. In addition, state 1|0|20 is in direct contact with state 18|9|13 and thus to the rest of the 18-mer cluster. A representative snapshot of the 18-mer at the end of the simulation indicates an elongated conformation rather than a globular one as observed for the 8-mer (Fig. 5). The central aggregation states for hexamer and heptamers are 6|9|15 and 7|7|16, respectively, which are preferentially formed from trimers. Dimers are directly connected to aggregation state 8|7|12 and thus to the other $N1 = 8$ states,

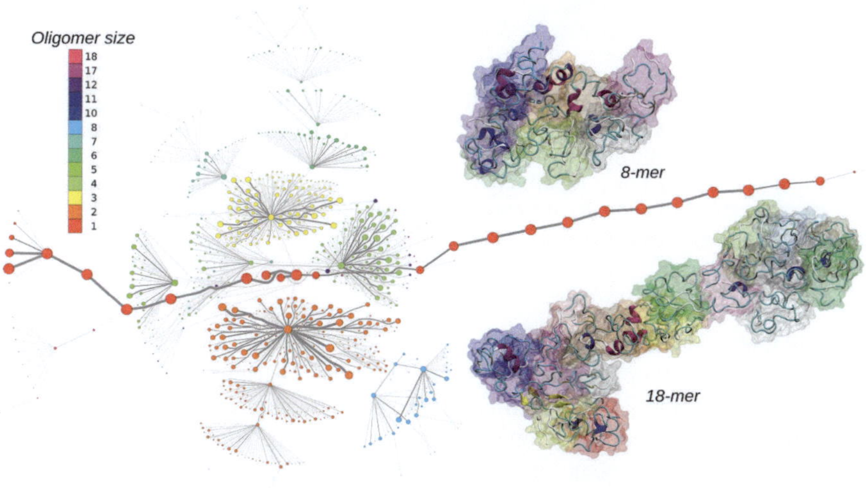

Fig. 5 Maximum flow transition network. The node area is proportional to the state population and the color corresponds to the oligomer size. The edge thickness correlates with the number of transitions. Two representative oligomer structures for the octamer and 18-mer are shown in cartoon and surface representation. Reproduced from [21] with permission from The Royal Society of Chemistry

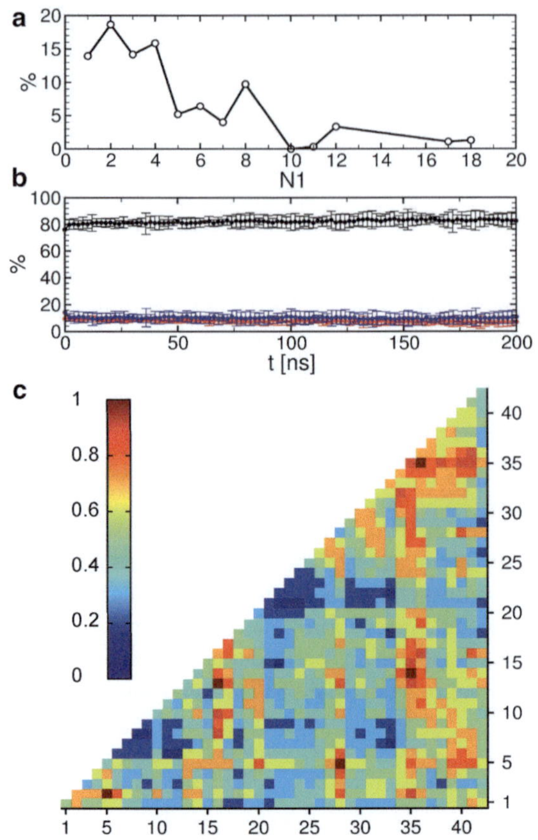

Fig. 6 Analysis of Aβ oligomers. (**a**) Relative oligomer mass distribution during aggregation. The probability was determined as the population of $N1$ multiplied by $N1$. (**b**) Secondary structure propensity as a function of time. *Black* corresponds to coil, *red* to β-strand, and *blue* to helix. (**c**) Inter-molecular contact map for all oligomers based on a cutoff distance of 0.75 nm. *Color coding* corresponds to the normalized number of contacts. Reproduced from [21] with permission from The Royal Society of Chemistry

indicating that octamers are largely formed by the addition of dimers to either tetramers or hexamers. We calculated the oligomer mass distribution (Fig. 6a), which revealed a higher population for dimers, tetramers, hexamers, octamers, 12-mers and 18-mers. Previous computational [7] and experimental studies [6] of Aβ aggregation report significant peaks in pentamers/hexamers and 12/13-mers, in agreement with our results.

Structural Changes To describe structural changes during the assembly process we calculated the time evolution of the secondary structure propensities (Fig. 6b). The initial β-strand propensity ($\sim 10\,\%$) decreased slightly throughout the simulation and had an average of 7.6±2.2 %. The average helical propensity was around

10.1±4.0 % and the average coil propensity started around 80 % and increased slightly, having an average value of 82.2±4.3 %. This indicates little change in the overall secondary structure. Experimental studies [6] indicated 13–20 % β-sheet for Aβ42 and Aβ40 while a more recent study shows a jump from ∼25 to ∼45 % when converting from monomers to tetramers of Aβ40 [63]. Another experimental study reported Aβ42 oligomers without β-sheet structure, in close agreement with our findings, which are on-pathway intermediates for fibril formation [64].

Interpeptide Interactions To investigate the key amino acids involved in the assembly process we calculated contact maps for the interface between any two proteins that are part of an oligomer (Fig. 6c). An important result from the intermolecular contact map is the proximity of hydrophobic regions from the C-terminus including L31–A42. In addition, region L33–V35 is in close proximity to region L17–F20. On the basis of solid-state NMR spectroscopy it has been shown that the C-terminal region is buried inside disc-shaped oligomers (pentamers and 10-mers) with strong contacts between F19 and L34 [64]. The 8-mer shown in Fig. 5 matches the description of the low-order oligomers with non-β-sheet secondary structure observed by Ahmed et al. [64]. Electrostatic interactions also contribute to oligomer formation by strong contacts formed by amino acid K16 with region S8–Q15 containing the negatively charged residue E11, and by K28 with region D1–D7 containing the negatively charged residues D1, E3, and D7.

2.2.3 Conclusions

Our results suggest that early Aβ aggregation precedes β-sheet formation. This is an important observation considering the low Aβ concentration in our study (∼0.8 mM) is close to experimental concentrations (∼30 μM–1 mM) [6] and smaller than concentrations from other computational studies (3.4 mM) [7]. At the same time, the structurally stable oligomers and the small changes in their secondary structure indicate a fast assembly process relative to the time scale corresponding to their conformational reorganization. Whether the oligomers observed here during a very early aggregation stage are similar to the toxic species observed in experiments is difficult to assess. While we found similarities with experimentally observed Aβ42 oligomers in terms of size and structure [7, 64], there is still a large debate regarding which Aβ oligomers are the toxic ones [65]. Some groups consider small oligomers with high content of β-sheet as toxic species [63], others suggest a second nucleation process where amyloid fibrils are present with small oligomer species as the source for toxic oligomers [66], and others propose that toxic Aβ oligomers have cross-β structure [67]. In future studies we plan to follow the further growth and structural conversion of the early oligomers and study their interactions with membranes. The aim should be that experiments probe at the same time the size, secondary structure and toxicity of low-order oligomers, allowing to directly relate simulation and experimental results.

Acknowledgements We gratefully acknowledge the computing time granted on the supercomputer JUROPA at Jülich Supercomputing Centre. We thank Prof. David J. Wales and Dr. Olujide O. Olubiyi for fruitful discussions.

References

1. Hardy J, Selkoe DJ (2002) Science 297(5580):353
2. Kirkitadze MD, Bitan G, Teplow DB (2002) J Neurosci Res 69(5):567
3. Silveira JR, Raymond GJ, Hughson AG, Race RE, Sim VL, Hayes SF, Caughey B (2005) Nature 437(7056):257
4. Haass C, Selkoe DJ (2007) Nat Rev Mol Cell Biol 8(2):101
5. Klein WL, Stine WB, Teplow DB (2004) Neurobiol Aging 25(5):569
6. Bitan G, Kirkitadze MD, Lomakin A, Vollers SS, Benedek GB, Teplow DB (2003) Proc Natl Acad Sci USA 100(1):330
7. Urbanc B, Betnel M, Cruz L, Bitan G, Teplow DB (2010) J Am Chem Soc 132:4266
8. Barz B, Urbanc B (2012) PLoS ONE 7(4):e34345
9. Melquiond A, Dong X, Mousseau N, Derreumaux P (2008) Curr Alzheimer Res 5(3):244
10. Sekijima M, Motono C, Yamasaki S, Kaneko K, Akiyama Y (2003) Biophys J 85(2):1176
11. Balbirnie M, Grothe R, Eisenberg DS (2001) Proc Natl Acad Sci USA 98(5):2375
12. Wales DJ (2010) Curr Opin Struc Biol 20(1):3
13. Strodel B, Whittleston CS, Wales DJ (2007) J Am Chem Soc 129(51):16005
14. Becker OM, Karplus M (1997) J Chem Phys 106(4):1495
15. Wales DJ, Miller MA, Walsh TR (1998) Nature 394(6695):758
16. Krivov SV, Karplus M (2004) Proc Natl Acad Sci USA 101(41):14766
17. Rao F, Karplus M (2010) Proc Natl Acad Sci 107(20):9152
18. Baba A, Komatsuzaki T (2007) Proc Natl Acad Sci USA 104(49):19297
19. Gfeller D, Rios PDL, Caflisch A, Rao F (2007) Proc Natl Acad Sci USA 104(6):1817
20. Barz B, Wales DJ, Strodel B (2014) J Phys Chem B 118(4):1003
21. Barz B, Olubiyi OO, Strodel B (2014) Chem Commun 50:5373
22. Riccardi L, Nguyen PH, Stock G (2012) J Chem Theory Comput 8(4):1471
23. Ford LR, Fulkerson DR (1956) Canad J Math 8(0):399
24. Krivov SV, Karplus M (2002) J Chem Phys 117(23):10894
25. Gomory RE, Hu TC (1961) J Soc Ind Appl Math 9(4):551
26. Nasica-Labouze J, Mousseau N (2012) PLoS Comput Biol 8(11):e1002782
27. Nasica-Labouze J, Meli M, Derreumaux P, Colombo G, Mousseau N (2011) PLoS Comput Biol 7(5):e1002051
28. Cheon M, Chang I, Hall CK (2011) Biophys J 101(10):2493
29. Osborne KL, Bachmann M, Strodel B (2013) Proteins 81(7):1141
30. Gsponer J, Haberthür U, Caflisch A (2003) Proc Natl Acad Sci USA 100(9):5154
31. Vitagliano L, Esposito L, Pedone C, De Simone A (2008) Biochem Biophys Res Commun 377(4):1036
32. Nguyen PH, Li MS, Stock G, Straub JE, Thirumalai D (2007) Proc Natl Acad Sci USA 104(1):111
33. Berendsen HJC, Postma JPM, Gunsteren WFV, Hermans J (1981) In: B. Pullman Intermolecular forces, no. 14 in The Jerusalem Symposia on Quantum Chemistry and Biochemistry (Springer, The Netherlands), pp 331–342
34. Hess B, Kutzner C, van der Spoel D, Lindahl E (2008) J Chem Theory Comput 4(3):435
35. Scott WRP, Hnenberger PH, Tironi IG, Mark AE, Billeter SR, Fennen J, Torda AE, Huber T, Krger P, van Gunsteren WF (1999) J Phys Chem A 103(19):3596
36. Feenstra KA, Hess B, Berendsen HJC (1999) J Comput Chem 20(8):786

37. Bussi G, Donadio D, Parrinello M (2007) J Chem Phys 126(1), 014101
38. Berendsen HJC, Postma JPM, van Gunsteren WF, DiNola A, Haak JR (1984) J Chem Phys 81(8):3684
39. Darden T, York D, Pedersen L (1993) J Chem Phys 98(12):10089
40. Essmann U, Perera L, Berkowitz ML, Darden T, Lee H, Pedersen LG (1995) J Chem Phys 103(19):8577
41. Klimov DK, Thirumalai D (2003) Structure 11(3):295
42. Frishman D, Argos P (1995) Proteins 23:566
43. Carter P, Andersen CA, Rost B (2003) Nucleic Acids Res 31:3293
44. Dima R, Thirumalai D (2002) Protein Sci 11(5):1036
45. Brandes U, Wagner D (2003) Graph drawing software. Springer, New York, pp 321–340
46. Humphrey W, Dalke A, Schulten K (1996) J Mol Graph 14(1):33
47. Colletier JP, Laganowsky A, Landau M, Zhao M, Soriaga AB, Goldschmidt L, Flot D, Cascio D, Sawaya MR, Eisenberg D (2011) Proc Natl Acad Sci USA 108(41):16938
48. Berhanu WM, Hansmann UHE (2012) Protein Sci 21(12):1837
49. Laganowsky A, Liu C, Sawaya MR, Whitelegge JP, Park J, Zhao M, Pensalfini A, Soriaga AB, Landau M, Teng PK, Cascio D, Glabe C, Eisenberg D (2012) Science 335(6073):1228
50. Liu C, Zhao M, Jiang L, Cheng PN, Park J, Sawaya MR, Pensalfini A, Gou D, Berk AJ, Glabe CG, Nowick J, Eisenberg D (2012) Proc Natl Acad Sci USA 109(51):20913
51. Berhanu WM, Hansmann UHE (2013) Proteins 81(9):1542
52. Xie L, Luo Y, Wei G (2013) J Phys Chem B 117(35):10149
53. Mehta AK, Lu K, Childers WS, Liang Y, Dublin SN, Dong J, Snyder JP, Pingali SV, Thiyagarajan P, Lynn DG (2008) J Am Chem Soc 130(30):9829
54. Reddy AS, Chopra M, de Pablo JJ (2010) Biophys J 98(6):1038
55. Hwang W, Zhang S, Kamm RD, Karplus M (2004) Proc Natl Acad Sci USA 101(35):12916
56. Meli M, Morra G, Colombo G (2008) Biophys J 94(11):4414
57. Zhang Z, Chen H, Bai H, Lai L (2007) Biophys J 93(5):1484
58. Berryman JT, Radford SE, Harris SA (2009) Biophys J 97(1):1
59. Park J, Kahng B, Hwang W (2009) PLoS Comput Biol 5(9):e1000492
60. Collins SR, Douglass A, Vale RD, Weissman JS (2004) PLoS Biol 2(10):e321
61. Narayanan S, Walter S, Reif B (2006) Chem BioChem 7(5):757
62. Klein WL, Krafft GA, Finch CE (2001) Trends Neurosci 24(4):219
63. Ono K, Condron MM, Teplow DB (2009) Proc Natl Acad Sci USA 106(35):14745
64. Ahmed M, Davis J, Aucoin D, Sato T, Ahuja S, Aimoto S, Elliott JI, Van Nostrand WE, Smith SO (2010) Nat Struct Mol Biol 17(5):561
65. Benilova I, Karran E, De Strooper B (2012) Nat Neurosci 15(3):349
66. Cohen SIA, Linse S, Luheshi LM, Hellstrand E, White DA, Rajah L, Otzen DE, Vendruscolo M, Dobson CM, Knowles TPJ (2013) Proc Natl Acad Sci USA 201218402
67. Stroud JC, Liu C, Teng PK, Eisenberg D (2012) Proc Natl Acad Sci USA 109(20):7717

Determination of the Size of the Primary and Secondary Folding Nuclei of Protofibrils from the Concentration Dependence of the Rate and the Lag-Time of Their Formation

Alexey V. Finkelstein, Nikita V. Dovidchenko, Olga M. Selivanova, Maria Yu. Suvorina, Alexey K. Surin, and Oxana V. Galzitskaya

Abstract A kinetic model of the process of formation of amyloid protofibrils is proposed, allowing calculation of the nucleus size solely from kinetic data. This model includes the primary nucleation, the linear growth of protofibrils binding monomers to their ends, and a possible exponential growth of protofibrils at the cost of either their fragmentation, or branching with the secondary nuclei, or growth from the surface. Theoretically, only the exponential growth is compatible with the existence of a pronounced lag-period (that can take much more time than the following protofibril growth itself). The obtained analytical solution allows us to determine the size of primary and secondary nuclei from experimentally obtained concentration dependences of the growth time of amyloid protofibrils and their lag-time/growth-time ratio. These theoretical results were used to analyze experimental data taken from the literature and our own experimental data on amyloid formation.

Keywords Amyloid protofibrils • Primary nucleus formation • Lag-time • Time of growth • Exponential growth • Bifurcation • Secondary nucleus formation • Fragmentation • Growth from the surface • Linear growth

---

A.V. Finkelstein (✉)
Institute of Protein Research, Russian Academy of Sciences, Pushchino, Moscow Region 142290, Russian Federation

Faculty of Biology, Moscow State Lomonosov University, 1-12 Leninskie Gory, Moscow 119991, Russian Federation
e-mail: afinkel@vega.protres.ru

N.V. Dovidchenko • O.M. Selivanova • M.Y. Suvorina • O.V. Galzitskaya
Institute of Protein Research, Russian Academy of Sciences, Pushchino, Moscow Region 142290, Russian Federation

A.K. Surin
Institute of Protein Research, Russian Academy of Sciences, Pushchino, Moscow Region 142290, Russian Federation

State Research Center for Applied Microbiology and Biotechnology, Obolensk, Serpukhov District, Moscow Region 142279, Russian Federation

L. Olivares-Quiroz et al. (eds.), *Physical Biology of Proteins and Peptides*,
DOI 10.1007/978-3-319-21687-4_3

Introduction

Filamentous aggregates are quite a ubiquitous phenomenon involved in many aspects of living matter—both normal (actin, tubulin) and pathogenic (hemoglobin S aggregates in sickle cell anemia, amyloid fibrils). However, the understanding of the rate laws that underlie the aggregation phenomenon is a challenging task despite the wealth of experimental data.

Oosawa et al. was the first to show that filament formation (in the case of actin) can be represented by a model of the so-called "linear growth" [1], which consists of two stages: primary nucleation and consecutive joining of monomers to the ends of a fibril. An analytical solution of the rate equations has shown that, in such a kinetic scheme, fibril mass accumulation during the initial aggregation phase (i.e., when only a small amount of aggregates exists) follows the quadratic dependence on time. Later, this model was generalized as described [2, 3] for different special cases. However, prediction of quadratic time-dependence of fibril mass accumulation during the initial phase seems to fail for hemoglobin S in sickle cell anemia [4], which led to the next major step in understanding of kinetic features on protein aggregation. To explain the observed effect of "extreme autocatalysis" and the strong concentration dependence, Ferrone et al. developed a model of "heterogeneous nucleation" [5]. This model assumes the following events: first, in the "normal" course of nucleation, a protofibril is formed, and then, on its surface, additional protofibrils can be formed. The concept of heterogeneous nucleation was an important contribution to the theory of protein aggregation. This model assumes that the surface capable of binding new monomers increases with increasing aggregates' size (because the surface of protofibrils can produce the secondary nuclei), which results in a comparatively long lag-period at the beginning of the process and a very rapid fibril mass accumulation ("exponential", rather than "quadratic" growth) after the lag-period.

Nowadays the interest to the protein aggregation kinetics is supported by extensive studies of amyloid formation. Amyloids are protein aggregates which usually have a fibrillar morphology with an intermolecular cross-beta structure, and they can be stained with special dyes (Congo red, ThT) [6]. Amyloids are associated with various pathologies, such as scrapie, mad-cow disease, Creutzfeldt–Jakob, Alzheimer's, Parkinson's diseases, and many other infectious, sporadic, and hereditary diseases [7]. Intricacy of mechanisms which drive the formation of amyloid aggregates should be addressed to the presence of signs of both linear and exponential growth scenarios.

Several mathematical models for polymerization reactions were proposed, based on a combination of the nucleation mechanism at the first stage of reaction and exponential aggregation at the second, where a potential candidate for the exponential growth mechanism is fragmentation [8, 9], in addition to "heterogeneous nucleation" [5]. Radford and coauthors approximated the kinetic behavior of β2-microglobulin aggregation using a "modular" system of analysis of the aggregation kinetics [9], which gave a possibility to choose a model in the best agreement with the experiment. Knowles and coauthors [10] analytically considered a model that included a stage of nucleation and a stage of exponential growth with a fragmentation scenario. These authors noted that the process of fragmentation plays a key role in approximation of experimental data. Also, the authors spotted that almost all (but for two) considered proteins, for which the kinetic data were approximated, showed linear scaling in coordinates {ln(T_{lag}) vs. ln[M]}, where M is monomer concentration and T_{lag} is duration of the lag period (i.e., the dependence of a kind $\ln\left(T_{lag}\right) = \text{const} + \gamma \ln\left[\text{M}\right]$, with constant $\gamma = -1/2$). In the same line, Cohen et al. [11] added the process of secondary nucleation (i.e., the ability to form a new protofibril nucleus on the surface of growing fibrils) to the kinetic scheme with the fragmentation mechanism [10] and solved the resultant equations analytically. To explain the scaling effect observed in [10], dependence of the secondary nucleus size on the lag-period was calculated. According to these calculation, γ is dependent on the secondary nucleus size n as $\gamma = -\left(n+1\right)/2$. In experiments with a substantial lag-period and $n = 0$ (i.e., in the absence of the secondary nucleus), one has the same solution as for the fragmentation process, so that the scaling in [10] was addressed to the specific mechanism of amyloid formation reaction. Morris et al. described a simple model [12] for the process of amyloid aggregation where exponential growth was modeled by including proportionality of the "ends" of the growing aggregate to the mass of the aggregate itself.

Despite the wide variety of scenarios and models used to explain characteristic kinetic features of the amyloid self-assembly process, there are still questions to answer. It is not always clear how individual stages influence the course of the reaction, because the analysis uses simplifications [10] or numerically simulates the kinetic scheme with a very large number of parameters [9]. Apparently, in nature one can meet all mechanisms of growth (Giehm and Otzen concluded that different conditions used to induce aggregation may lead to different fibrillization pathways [13]). In particular, for prions (PrP), it is believed that fragmentation is the main factor of their spontaneous fast growth after a long lag-period [14]. For insulin, it has been shown that aggregation occurs by the lateral growth scenario, which is one of the types of branching [15]. Branching in itself (when growing fibrils have a tree-like structure) was directly shown to be present in amyloidogenesis of glucagon [16] and Aβ-peptide [17, 18]. Recent experiments [19] have shown β-structural droplets that may result from the amyloid growth from the surface; still, this requires a deeper analysis.

Generalized models of different scenarios of growth of protofibrils have been considered in our papers [20, 21]. Now we give a brief overview of the developed theory and compare its results with experiments on some amyloid-forming proteins.

Following Oosawa et al. [1] and Collins et al. [22], we assume that the growth of aggregates (for all the scenarios considered) proceeds by adsorption of monomers floating in a rather diluted solution. Concentration of these monomers is the main parameter of our theory. For simplicity, we assume that a "monomer" includes only one protein molecule; however, there exists an evident possibility of generalization of the same approach for the case when a "monomer" consists of several protein molecules floating together. Despite the fact that such a model is simplified, and various scenarios of growth of the aggregates have been already analyzed in a number of

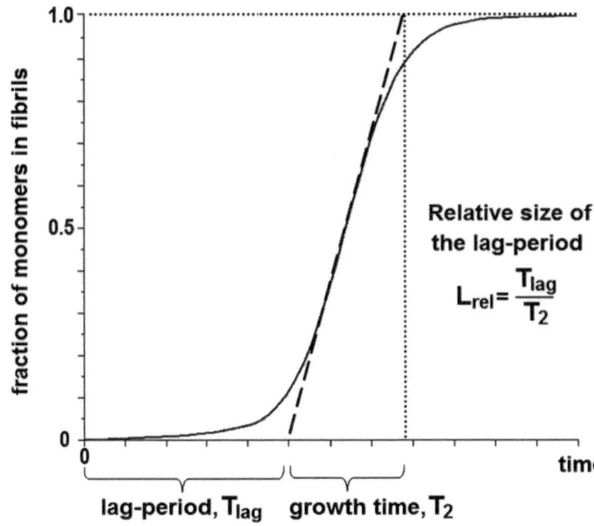

Fig. 1 Visual representation of the lag-period, transition (growth) time, and the relative size L_{rel} of the lag-period obtained from the given kinetic curve. In the given example, $L_{rel} \approx 1.8$

papers [1, 9, 10, 12], we outlined some novel important relations existing between experimentally measurable parameters. This is, in particular, the *relative* duration of the lag-time, L_{rel}, which is the ratio of the time of lag-period T_{lag} to the transition time T_2 (Fig. 1). This value and its concentration dependence reveal some specific features of kinetic mechanisms of amyloid proteofibril growth [20] (see below).

It should be noted that not all scenarios of exponential growth (fragmentation, branching, growth from the surface) can be distinguished from one another on the basis of kinetic experiments alone. This statement implies the requirement for additional, non-kinetic experimental data, which, if combined with kinetic data, can indicate which scenario of growth takes place in a particular case. However, the suggested model enables one to calculate, directly from experimental kinetic data, the number of monomers in the primary folding nucleus (this "nucleus" is the most unstable aggregate on the protofibril formation pathway that is a predecessor of amyloidal seeds—the smallest stable and thus directly observable aggregates), as well as the number of monomers in the secondary folding nucleus that leads to the exponential growth with branching. The information on the size of nuclei is important for determining the mechanism of the amyloid formation in each particular case.

Theory

Model of Linear Growth

The simplest model describes formation of amyloid protofibrils as a straight, successive and irreversible (after nucleus formation) reaction, where the subsequent growth is initiated by the ends of polymers (Fig. 2).

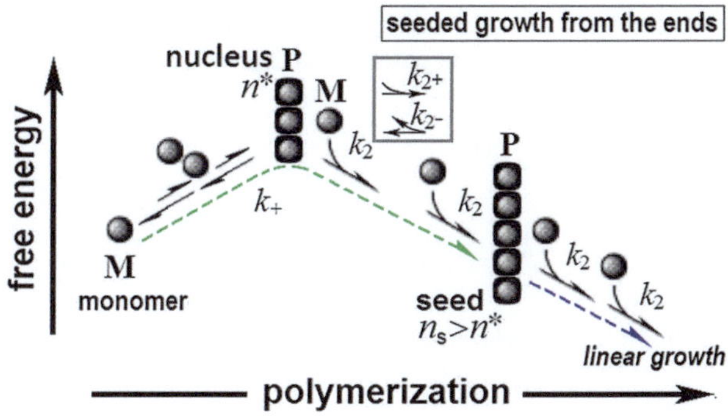

Fig. 2 Free energy change during the initiation and linear growth of a protofibril. *Ball*: free monomer (**M**) in solution, i.e., in a non-polymerized form. *Square*: monomer in a polymer. "Nucleus" (of n^* monomers) is the most unstable aggregate on the reaction pathway; "polymers" (**P**) are linear aggregates. "Seed" is the smallest stable (against separate monomers) polymer. k_+ is the effective rate constant of a many-particle reaction of formation of the seed; k_+ depends on concentration [M] of free monomers in solution. k_2 is the effective, concentration-dependent rate constant for polymer elongation: $k_2[\mathrm{M}] = k_{2+}[\mathrm{M}] - k_{2-}$, where k_{2+} is the rate constant for polymer elongation by one monomer, and k_{2-} is the rate constant for polymer shortening by one monomer. Assuming a significant excess of free monomers in solution, $k_2 \approx k_{2+}$

Such a model has been introduced by Oosawa et al. [1], and an analytical solution for the appropriate system of kinetic equations was provided. However, it is useful to rewrite the equations in terms of characteristic times (Fig. 1) to obtain the key values for this (linear growth) and the other (exponential growth) scenarios in a convenient unified form. The solution of these kinetic equations is based on the transition state theory [23, 24].

The linear growth process can be divided into two kinetic stages:

1. Formation of the polymer seed from a few monomers: $n_S\mathbf{M} \xrightarrow{k_+} \mathbf{P}_{n_S}$ (the reverse process $\mathbf{P}_{n_s} \to n_s\mathbf{M}$ plays no significant role because concentration of seeds, which are rapidly involved in further polymerization, is low: it is close to concentration of the unstable nuclei themselves [25]). The seed consists of n_s monomers M; n_s is the smallest number of monomers in the *stable* polymer **P**. The process goes on with an effective rate constant k_+ that is defined by the height of the free-energy barrier separating monomer and polymer states. This barrier corresponds to the nucleus comprising n^* monomers; $n^* < n_s$ (see Fig. 2); therefore, the height of the free-energy barrier is proportional to the [concentration of monomers] in the power n^*.
2. The process $\mathbf{P} + \mathbf{M} \xrightarrow{2\text{ends}\times k_2} \mathbf{P}$ (more precisely, the sum of all $\mathbf{P}_n + \mathbf{M} \xrightarrow{2\text{ends}\times k_2} \mathbf{P}_{n+1}$ with $n \geq n_s$) describes further polymerization of monomers **M** at the ends of already formed protofibrils **P**; the multiplier 2 is introduced because concentration of the ends is twice higher than that of polymers **P**; k_2 is the rate constant of polymerization (see legend to Fig. 2). The reverse process $\mathbf{P}_n + \mathbf{M} \leftarrow \mathbf{P}_{n+1}$ does not play a role when the process is irreversible.

These two stages are described by the following kinetic equations with respect to monomer concentration [M] and concentration of the ends of all stable polymers [P]:

$$\begin{cases} \dfrac{d[\mathrm{P}]}{dt} = 2k_+[\mathrm{M}]^{n^*} \\ \dfrac{d[\mathrm{M}]}{dt} = -k_+ n_s [\mathrm{M}]^{n^*} - k_2[\mathrm{P}][\mathrm{M}] \approx -k_2[\mathrm{P}][\mathrm{M}]. \end{cases} \tag{1}$$

In the latter equation of this system (first obtained by Oosawa et al. [1]) one can neglect the small term $-k_+ n_s [\mathrm{M}]^{n^*}$ (i.e., monomers spent to form a seed) because the fraction of monomers involved in the seeds is much smaller than the fraction of monomers involved in all the nascent polymers. At the time $t=0$, $[\mathrm{P}]_{t=0}=0$; thus, $\left.\dfrac{d[\mathrm{M}]}{dt}\right|_{t=0} = 0$,

but $\left.\dfrac{d^2[\mathrm{M}]}{dt^2}\right|_{t=0} = -k_2 \left.\dfrac{d[\mathrm{P}]}{dt}\right|_{t=0} [\mathrm{M}]_{t=0} - k_2[\mathrm{P}]_{t=0} \left.\dfrac{d[\mathrm{M}]}{dt}\right|_{t=0} = -2k_2k_+[\mathrm{M}_\Sigma]^{n^*}$, where $[\mathrm{M}_\Sigma] \equiv [\mathrm{M}]_{t=0}$ is the total concentration of free monomers at $t=0$, i.e., before the polymerization reaction. Denoting by $\mu(t) \equiv 1 - \dfrac{[\mathrm{M}]_t}{[\mathrm{M}_\Sigma]}$ a fraction of aggregated monomers at the time t, and introducing the characteristic time

$$T_{init} = \sqrt{1/\left(2k_2k_+[\mathrm{M}_\Sigma]^{n^*}\right)} \tag{2}$$

that scales the beginning of the process, one has $\mu(t) \cong \dfrac{1}{2}\left(\dfrac{t}{T_{init}}\right)^2$; this is the "quadratic growth" [1].

Actually, System (1) can be strictly solved for the whole process from the equation $\dfrac{d^2}{dt^2}\ln\left(\dfrac{[\mathrm{M}]_t^{n^*}}{[\mathrm{M}_£]^{n^*}}\right) = -2k_2k_+[\mathrm{M}_£]^{n^*}$ (see item 1 in Supplementary material to [20], or http://phys.protres.ru/resources/AMYL/P1_suppl_corr.pdf, where a few mis prints in the above mentioned Supplementary material have been corrected). The result, for the "linear growth", is

$$\mu(t) = 1 - \frac{4^{1/n^*}\exp(-t/\mathrm{T})}{\left[1+\exp(-n^*t/\mathrm{T})\right]^{2/n^*}} \tag{3}$$

where

$$\mathrm{T} = \sqrt{n^*/\left(4k_2k_+[\mathrm{M}_\Sigma]^{n^*}\right)} = \sqrt{n^*/2}\cdot T_{init} \tag{4}$$

is the characteristic time of linear growth of $\mu(t)$ at $t \to \infty$, i.e., at the end of the polymerization reaction. $\mu(t)$ is a monotonic S-shaped function spanning from $\mu(0)=0$ to $\mu(\infty)=1$.

Table 1 Theoretical dependencies of parameters of experimental kinetic curves on the total concentration of monomers and the nucleus sizes

Scenario of growth	Dependence of the characteristic duration of the aggregate growth time (T_2) and of the relative lag-period duration (L_{rel}) on the total concentration of monomers $[M_\Sigma]$	Calculation of sizes of the primary (n^*) and secondary (n_2) nuclei from the experimental kinetic data alone
Linear growth	$\ln T_2 = const - \frac{n^*}{2}\ln[M_\Sigma]$	$n^* = -2\frac{d(lnT_2)}{d(\ln[M_\Sigma])}$
	$L_{rel} = f_L(n^*) < 0.2$	
Growth from the surface	$\ln T_2 = const - \ln[M_\Sigma]$	$n^* = 2 - 4\frac{d(L_{rel})}{d(\ln[M_\Sigma])}$
	$L_{rel} = const' - \frac{(n^*-2)}{4}\cdot\ln[M_\Sigma]$	
Growth with fragmentation	$\ln T_2 = const - \frac{1}{2}\ln[M_\pounds]$	$n^* = 1 - \frac{d(L_{rel})}{d(\ln[M_\Sigma])}$
	$L_{rel} = const' - (n^*-1)\cdot\ln[M_\Sigma]$	
Growth with bifurcation	$\ln T_2 = const - \frac{n_2+1}{2}\ln[M_\Sigma]$	$n_2 = -1 - 2\frac{d(\ln T_2)}{d(\ln[M_\Sigma])}$
	$L_{rel} = const' - (n^*-n_2-1)\cdot\ln[M_\Sigma]$	$n^* = 1 + n_2 - \frac{d(L_{rel})}{d(\ln[M_\Sigma])}$

The characteristic time of such a protofibril growth (determined as T2= [max(dμ/dt)]-1, see Fig. 1) can be presented [20] as

$$T_2 = T_{init}\cdot f_T(n^*) = [M_\Sigma]^{-n^*/2}\cdot\left(f_T(n^*)/\sqrt{2k_2k_+}\right), \tag{5}$$

and the relative lag-period duration (determined as shown in Fig. 1) is

$$L_{rel} = f_L(n^*), \tag{6}$$

where the functions $f_T(n^*)$,$f_L(n^*)$ depend on the nucleus size n^* *only* and are independent of $[M_\Sigma]$ and the rate constants. Thus, L_{rel} is independent of $[M_\Sigma]$, while durations of all stages of the linear growth are proportional to $[M_\Sigma]^{-n^*/2}$ (see item 1 in Supplementary material to [20], equations (S.1.6)–(S.1.10), or http://phys.protres.ru/resources/AMYL/P1_suppl_corr.pdf, and Table 1 below). The computed $L_{rel}(n^*)$ function (presented in Fig. 3 of [20]) gradually and monotonically decreases from ≈0.17 at $n^*=1$ to ≈0.1 at $n^*=5$, ≈0.07 at $n^*=10$, and so on, down to 0 at $n^*\to\infty$.

Thus, the experimental values $L_{rel}>0.2$ cannot be described by the model of linear growth.

However, experimental data reveal $L_{rel}>0.2$ in many cases [4, 25–28], which means that the linear growth model is not applicable for them, and one needs to introduce scenarios where the number of protofibrils increases exponentially [5, 8–11].

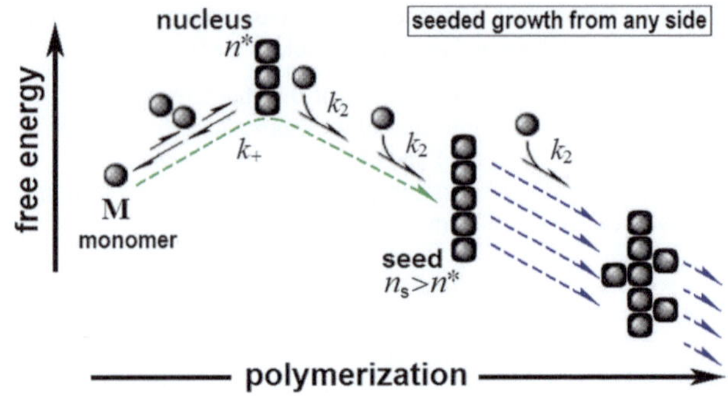

Fig. 3 Irreversible aggregation where growth occurs at any site of the surface

The Extreme Model of Exponential Growth: The "Growth from the Surface" Scenario

Figure 3 shows a reaction that is initiated by some seed with subsequent monomer bonding to any site of the aggregate. It should be noted that in this scenario polymer ends play no special role, and therefore, fibrils are not the favorable structures; rather, droplet-like structures will be formed. However, because some experiments show certain signs of this scenario (non-fibrillar β-structure-rich aggregates) [19], this case is of potential interest as an extreme and the simplest model of the exponential growth. A similar scheme was mentioned previously by Morris et al. [12]; however, no detailed analysis has been done.

In the approximation given by the steady-state theory [23, 24], an irreversible aggregation of this kind is described by two kinetic phases:

1. Formation of the seed from a few monomers: $n_S\mathbf{M} \underset{k_+}{\rightarrow} \mathbf{P}_{n_S}$ (as in the "linear growth" scenario).
2. The sum of all subsequent bindings ($\mathbf{P}_n + \mathbf{M} \underset{n \times k_2}{\rightarrow} \mathbf{P}_{n+1}$, with $n \geq n_s$) of monomers to the aggregate; the reverse process $\mathbf{P}_\mathrm{n} + \mathbf{M} \leftarrow \mathbf{P}_{\mathrm{n}+1}$ plays no role when the reaction is irreversible. The speed of the process is proportional to the number $[\mathrm{P}]_\mathrm{all} = [\mathrm{M}_£] - [\mathrm{M}]$ of all monomers in all aggregates.

As a result, we obtain:

$$\frac{d[\mathrm{M}]}{dt} = -2k_+ n_s [\mathrm{M}]^{n^*} - k_2\left([\mathrm{M}_\Sigma] - [\mathrm{M}]\right)\cdot[\mathrm{M}]. \tag{7}$$

Equation (7) can be easily solved for early stages of the reaction, when $[\mathrm{M}_\Sigma] - [\mathrm{M}] \ll [\mathrm{M}_\Sigma]$ (see Supplementary material to [20] or http://phys.protres.ru/resources/AMYL/P1_suppl_corr.pdf, item 2). The result is a monotonic S-shaped function $\mu(t)$ spanning from $\mu(0) = 0$ to $\mu(\infty) = 1$:

$$\mu(t) = \frac{1 - e^{-\hat{\varkappa}t}}{1 + \hat{\lambda}\cdot e^{-\hat{\varkappa}t}}, \tag{8}$$

where $\widehat{x} = k_2\left[M_\Sigma\right]\left(1+1/\widehat{\lambda}\right)$ and $\widehat{\lambda} = \dfrac{k_2}{2k_+ n_s \left[M_\Sigma\right]^{n^*-2}}$. It should be noted that $\lambda \gg 1$ in the most interesting case when the rate of the n^*-molecular (see Fig. 3) initiation reaction, connected with k_+, is much smaller than the rate of the further bimolecular growth reaction, connected with k_2. In this case $\mu(t) \approx \left(e^{\widehat{x}t} - 1\right)/\widehat{\lambda}$ at early stages of the reaction, i.e., when $e^{\widehat{x}t} \ll \widehat{\lambda}$. The duration of the exponential growth period is

$$T_2 = \left[\max\frac{d\mu}{dt}\right]^{-1} = \frac{4}{\left(1+1/\widehat{\lambda}\right)\widehat{x}} = \frac{4}{k_2\left[M_\Sigma\right]}. \tag{9}$$

An exponential growth increases the value of L_{rel} (as illustrated by Fig. 5 below). This value can be easily found when $\widehat{\lambda} \gg 1$ (actually, when $\widehat{\lambda} > 20$):

$$L_{rel} \approx \frac{1}{4}\left(\ln\widehat{\lambda} - 2\right) = \frac{1}{4}\left(\ln\left\{\frac{k_2}{2k_+ n_s \left[M_v\right]^{n^*-2}}\right\} - 2\right). \tag{10}$$

(see http://phys.protres.ru/resources/AMYL/P1_suppl_corr.pdf or Supplementary material to [20], item 2).

The important conclusion that follows is that there is no upper limit of L_{rel} (as it was in the case of the linear regime of growth, see Fig. 3) when $[M_\Sigma] \rightarrow 0$ and $n^* > 2$, as well as when $[M_\Sigma] \rightarrow \infty$ and $n^* < 2$, while L_{rel} does not depend on $[M_\Sigma]$ at $n^* = 2$ (see Table 1 below). A similar conclusion (as it will be shown below) is applicable to other forms of exponential regimes.

"Fragmentation" Scenario: Another Model of Exponential Growth

This model, also describing formation of amyloids as irreversible polymerization, is shown in Fig. 4. This scheme was first used for description of kinetics of actin filament formation [29]; an analytical solution was presented by Knowles et al. [10] in 2009. Because of considering "secondary nucleation" and "fragmentation" cases together, the authors' analysis lacked some major relations, which could be used for determination of the nucleus size, n^*; therefore, in this paper we will reconsider the both cases separately.

In the approximation given by the transition state theory [23, 24], a quasi-stationary process of irreversible polymerization with fragmentation is described by three kinetic phases:

1. Formation of the seed: $n_S \mathbf{M} \xrightarrow[k_+]{} \mathbf{P}_{n_S}$ (as in all the earlier scenarios).
2. Growth from the ends: $\mathbf{P}_n + \mathbf{M} \xrightarrow[2\_\text{ends} \times k_2]{} \mathbf{P}_{n+1}$ (as in all the earlier scenarios).
3. Fragmentation $\mathbf{P} \underset{l_P \times \lambda_+}{\overset{\lambda_-}{\longleftrightarrow}} \mathbf{P} + \mathbf{P}'$ of protofibrils: a protofibril divides into two stable

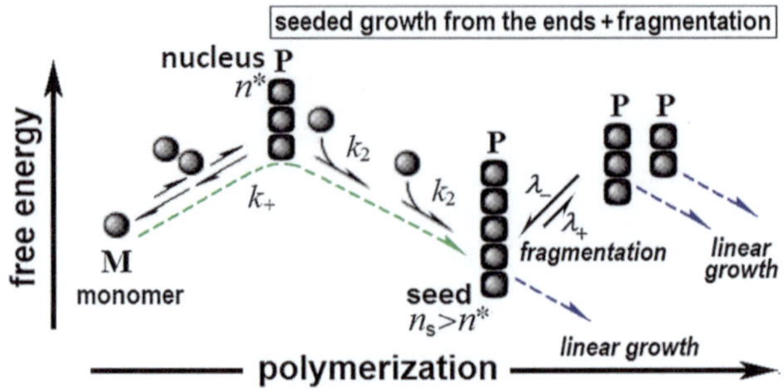

Fig. 4 Dependence of the free energy on the reaction coordinate in the process of amyloidogenesis with fragmentation. The rate constant λ_- is similar to the rate constant k_{2+} in Fig. 2 (and λ_+ is similar to k_{2-}), but $\lambda_- \neq k_{2+}$ (and $\lambda_+ \neq k_{2-}$) because fragmentation is not connected with reorganization of monomers at the new fibril ends, while a single monomer that dissociates form the fibril end acquires a form inherent to the free monomer **M** in solution

parts; the rate of fragmentation of a single protofibrill is roughly proportional to its length, l_p, with the rate constant λ_+. The possibility to merge two protofibrils into one (with the rate constant λ_-) is not important at a low concentration of protofibrils at the beginning of the process.

These stages are described by the following equations with respect to the concentration [M] of monomers and the concentration [P] of ends of all the stable polymers **P**:

$$\begin{cases} \dfrac{d[\mathrm{P}]}{dt} = 2\mathrm{k}_+ [\mathrm{M}]^{n^*} + 2\lambda_+ \cdot ([\mathrm{M}_\Sigma] - [\mathrm{M}]) - \lambda_- \cdot [\mathrm{P}]^2 \approx \\ \qquad\qquad 2k_+ [\mathrm{M}_\Sigma]^{n^*} + 2\lambda_+ ([\mathrm{M}_\Sigma] - [\mathrm{M}]) \\ \dfrac{d[\mathrm{M}]}{dt} = -k_+ n_s [\mathrm{M}]^{n^*} - k_2 [\mathrm{P}][\mathrm{M}] \approx -k_2 [\mathrm{P}][\mathrm{M}_\Sigma] \end{cases} \tag{11}$$

In the second equation of this system (cf. Ferrone and co-authors [5, 29] and Wegner [30]) one neglects (as in System (1)) the small term $-k_+ n_s [\mathrm{M}]^{n^*}$; also, at the beginning of the process, $[\mathrm{M}_\Sigma] - [\mathrm{M}] \ll [\mathrm{M}_\Sigma]$, and [P] is small; thus, the term $\lambda_- [\mathrm{P}]^2$ can be neglected in the first equation, and [M] can be replaced by $[\mathrm{M}_\Sigma]$ in $\mathrm{k}_+ [\mathrm{M}]^{n^*}$ and in k_2[P][M]. Then, for early stages of the reaction,

$$\mu(t) \equiv 1 - \frac{[M]_t}{[M_\Sigma]} = \frac{1}{2}\left(\frac{\mathrm{T}_2}{T_{init}}\right)^2 \left(e^{t/\mathrm{T}_2} + e^{-t/\mathrm{T}_2} - 2\right), \tag{12}$$

where T_{init} is defined by (2), and T_2 is the characteristic time for exponential growth with fragmentation:

$$\mathrm{T}_2 = \sqrt{1/(2\lambda_+ k_2 [\mathrm{M}_\Sigma])} = T_{init} \sqrt{k_+ [\mathrm{M}_\Sigma]^{n^*-1} / \lambda_+}, \tag{13}$$

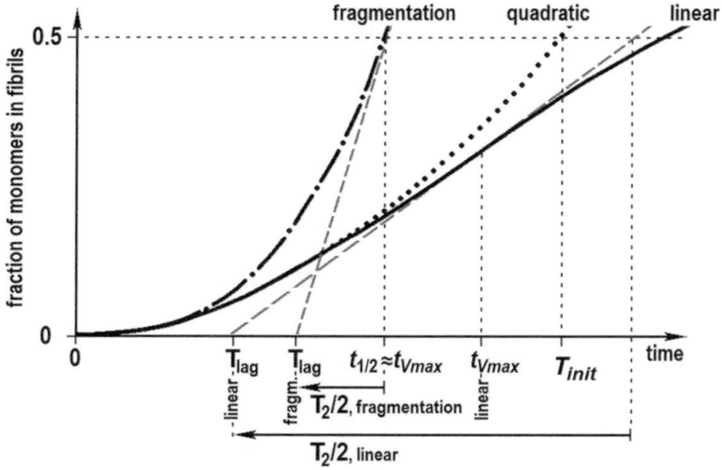

Fig. 5 A scheme illustrating comparison of the characteristic times and lag-periods for two amyloid-growth scenarios: the "linear growth" with $n^*=1$ (continuous line) and "growth with fragmentation" with $T_2/T_{init}=1/4$ (*dash-dotted line*). $T_{init}=\left(2k_2k_+\left[M_\Sigma\right]^{n^*}\right)^{-1/2}$, with all its parameters, is the same for the both cases. The *gray dashed lines* correspond to the maximal slopes of the curves in the both cases. $L_{rel}\approx 0.2$ in the first case and ≈ 2 in the last one. The *thick dotted curve* shows the "quadratic growth"; as seen, it describes the very beginning of the exponential growth

(see item 3 in http://phys.protres.ru/resources/AMYL/P1_suppl_corr.pdf or in Supplementary material to [20]). From this, $\mu(t)\approx\frac{1}{2}\left(\frac{t}{T_{init}}\right)^2$, as in the "linear growth" at small $t<T_2$; here, fragmentation plays no role. However, when $T_2 \ll T_{init}$ (i.e., when the fragmentation rate λ_+ is much higher than the fibril initiation rate $k_+\left[M_\Sigma\right]^{n^*-1}$) we see an exponential growth that achieves $\mu\approx 1/2$ at the time

$$t_{1/2}\approx T_2\ln\left(\frac{T_{init}}{T_2}\right)^2=T_{init}\sqrt{\frac{k_+\left[M_\Sigma\right]^{n^*-1}}{\lambda_+}}\ln\left(\frac{\lambda_+}{k_+\left[M_\Sigma\right]^{n^*-1}}\right).$$

It should be noted that the values $\mu\left(t>t_{1/2}+T_2/2\right)$ given by (12) approach unity and then exceed unity, but this continuation has no physical sense, being a result of replacement of $k_2[P][M]$ by $k_2[P][M_\Sigma]$ in (11).

Figure 5 (which, for the above mentioned reason, is limited to the region where $\mu(t)\le 1/2$) illustrates comparison of curves for the "linear" and "with fragmentation" scenarios.

Provided $T_2 \ll T_{init}$, we have $T_2 \ll t_{1/2} \ll T_{init}$, and the "relative lag-time duration" is

$$L_{rel}=\frac{t_{1/2}-T_2/2}{T_2}\approx\ln\left(\frac{T_{init}}{T_2}\right)^2-\frac{1}{2}\equiv\ln\left(\frac{\lambda_+}{k_+\left[M_\Sigma\right]^{n^*-1}}\right)-\frac{1}{2}. \quad (14)$$

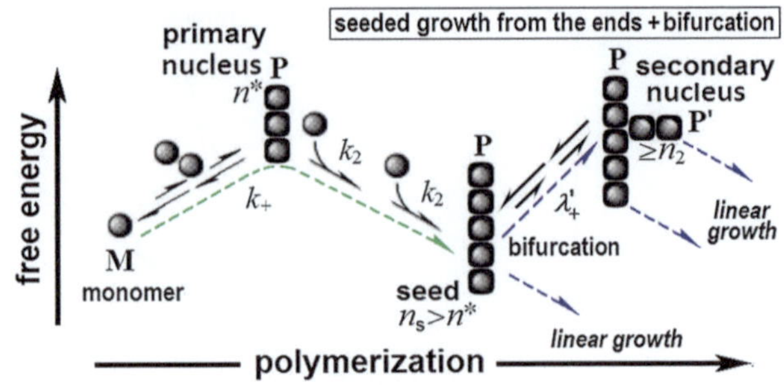

Fig. 6 Dependence of free energy on the reaction coordinate in amyloidogenesis with bifurcation. n_2 is the number of monomers in the nucleus of branching (i.e., in the "secondary nucleus")

Figure 5 and (6) and (12) show that L_{rel} for the fragmentation process (as for any exponential growth process) is greater than L_{rel} for the linear growth process with the same nucleus size n^*.

The dependence of L_{rel} on $[M_\Sigma]$ vanishes when $n^*=1$ (see Table 1 below). In the contrast to the "growth from the surface" scenario, where L_{rel} decreases with increasing $[M_\Sigma]$ when $n^*>2$, in the "fragmentation scenario" L_{rel} decreases with increasing $[M_\Sigma]$ when $n^*>1$, and the case where L_{rel} increases with increasing $[M_\Sigma]$ is absent (because n^* cannot be less than 1).

"Bifurcation" Scenario: One More Model of Exponential Growth

Figure 6 presents a model of formation of amyloids that includes a branching reaction [5, 11]. The branching, or bifurcation, is initiated by a secondary nucleus attached to some inner part of the polymer **P**; further growth occurs at all ends of the linear polymer and its branches. Initially, such a model has been applied to hemoglobin aggregation during sickle cell anemia decease [4, 5]. An analytical solution was provided (as in the fragmentation scenario) in [10, 11]; still, as mentioned above, the authors' analysis lacks some major relations; therefore, here we have reconsidered [20] equations that follow from the branching scheme.

The "polymerization with branching" process is described by three kinetic phases.

1. Formation of the seed: $n_S\mathbf{M} \underset{k_+}{\rightarrow} \mathbf{P}_{n_S}$ (as in all the earlier scenarios).
2. Growth from the ends: $\mathbf{P}_n + \mathbf{M} \xrightarrow[\text{all\_ends}\times k_2]{} \mathbf{P}_{n+1}$ (as in all the earlier scenarios).
3. Emergence of a stable branch on the first protofibril: $\mathbf{P} \underset{l_p \times \lambda_+'}{\rightarrow} \mathbf{P}, \mathbf{P}'$; the rate of branching of a single protofibril is roughly proportional to its length (l_p) with the rate constant $\lambda_+' \left[\mathrm{M}_£\right]^{n_2}$ where n_2 is the number of monomers in the secondary nucleus (i.e., the nucleus of branching). Unlike the fragmentation process shown in Fig. 4 which adds to the system two points of growth (two ends of a new protofibril), branching adds only one point of growth (the end of a branch P′).

From the mathematical point of view, this is the only difference between equations that describe bifurcation and fragmentation. Namely, the substitution

$2\lambda_+ \Rightarrow \lambda'_+ [\mathrm{M}_\Sigma]^{n_2}$ converts (11) and the following ones that describe amyloidogenesis with fragmentation into equations that describe amyloidogenesis with bifurcation.

In particular, the characteristic time for exponential growth via branching is

$$\mathrm{T}_2 = \sqrt{1/\left(\lambda'_+ k_2 [\mathrm{M}_\Sigma]^{n_2+1}\right)} = T_{init}\sqrt{2k_+ [\mathrm{M}_\Sigma]^{n^*-n_2-1} / \lambda'_+}, \quad (15)$$

and the "relative lag-time duration" (when $\mathrm{T}_2 \ll T_{init}$, i.e., when $\lambda'_+ \gg 2k_+ [\mathrm{M}_\Sigma]^{n^*-n_2-1}$) is

$$\mathrm{L}_{\mathrm{rel}} = \ln\left(\lambda'_+ / \left(2k_+ [\mathrm{M}_\Sigma]^{n^*-n_2-1}\right)\right) - \tfrac{1}{2}. \quad (16)$$

The dependence of $\mathrm{L}_{\mathrm{rel}}$ on $[\mathrm{M}_\Sigma]$ vanishes when $n^* - n_2 = 1$, and $\mathrm{L}_{\mathrm{rel}}$ increases with increasing $[\mathrm{M}_\Sigma]$ when $n^* - n_2 < 1$ and decreases when $n^* - n_2 > 1$.

It should be noted that the case of $n_2 = 0$ is quite possible: this case means that the initiation energy is connected with deformation of the protofibril rather than with deformation of the adhering monomer. This case was considered in our work [21].

The End of the Polymerization Process

While the rate of amyloid growth is limited mostly by a low concentration of already initiated fibrils at the beginning of amyloidogenesis, at the end of this process this rate is limited by concentration of free monomers (actually, this rate can also be effected by interactions of already formed protofibrils, but this complicated phenomenon will not be considered here).

The concentration [M] of free monomers is "low" when the rate of their attachment to the fibril ($k_{2+}[\mathrm{M}]$, see Fig. 2) becomes close to the rate of their separation from the fibril (k_{2-}); at the critical concentration $[\mathrm{M}_0] = k_{2-} / k_{2+}$ fibrils stop to grow. When [M] approaches $[\mathrm{M}_0]$, one should use $k_{2+}[\mathrm{M}] - k_{2-}$ instead of $k_2[\mathrm{M}]$ or $k_2[\mathrm{M}_\Sigma]$ in (1), (7) and (11). On the other hand, concentration of points capable bind new monomer can be considered as a constant, $[\mathrm{P}_0]$. Thus, the equation

$$\frac{d[\mathrm{M}]}{dt} \approx -k_{2+}[\mathrm{P}_0]\left([\mathrm{M}] - [\mathrm{M}_0]\right) \quad (17)$$

describes the end of any polymerization process.

$$\left[\mathrm{M}(t > t_0)\right] = [\mathrm{M}_0] + \left\{\left[\mathrm{M}(t_0)\right] - [\mathrm{M}_0]\right\}\exp\left\{-k_{2+}[\mathrm{P}_0](t - t_0)\right\}. \quad (18)$$

is its solution at large times $t > t_0$; $[\mathrm{M}(t_0)]$ can be estimated as ≈2[M0], assuming that $[\mathrm{M0}] \ll [\mathrm{M}_\Sigma]$. It is noteworthy that a dependence of this kind should be generally typical for amyloidogenesis where *any* kind of formation of new seeds, fibrils or branches is absent or negligible (for example, when seeding is made by a preliminary agitation and especially by preformed fibrils [26], like in the case of prion replication [31]).

Results and Discussion

Summary of the Theoretical Estimates

The obtained dependencies of $\ln T_2$ and L_{rel} on $\ln[M_\Sigma]$ are summarized in Table 1.

These dependencies can be used to estimate the mechanism (or a couple of possible mechanisms) of amyloid formation from a set of experiments where only the value of $[M_\Sigma]$ is varied. This estimate can be done rather easily, because the predicted [20] dependencies of L_{rel} and $\ln T_2$ on $\ln[M_\Sigma]$ are linear (while that of $\ln T_{lag}$ on $\ln[M_\Sigma]$, used in [10], is not: see Table 1 in [20]). Also, these dependencies provide a possibility to estimate the sizes n^*, n_2 of the primary and secondary nuclei from kinetic experimental data presented in coordinates ($\ln T_2$, $\ln[M_\Sigma]$) and (L_{rel}, $\ln[M_\Sigma]$).

For example, if $L_{rel}>0.2$, the linear growth model is excluded, and, independently of $\ln T_2$ and L_{rel} dependence on $\ln[M_\Sigma]$, only the exponential growth scenario remains. The dependences of $\ln T_2$ on $\ln[M_\Sigma]$ and L_{rel} on $\ln[M_\Sigma]$ are formally different (see Table 1) for different exponential scenarios (i.e., "growth from the surface", "fragmentation" and "bifurcation"). However, it is impossible, e.g., to distinguish the bifurcation from fragmentation scenario from kinetics *alone* if the size n_2 of the secondary nucleus of bifurcation is zero; this implies the need for direct observations of fibril shapes to distinguish these cases (see [20, 21] for more details).

Figures 7 and 8 illustrate determination of the fibrillation scenario and the nucleus sizes (n^*, n_2) for one protein—human insulin [21].

The growth of amyloid fibrils was studied [21] (Fig. 7) by thioflavin T (ThT) fluorescence, which drastically increases when ThT is bound to fibrils [32]. Fibril formation has been also followed [21] by electron microscopy (EM), as well as by fluorescent microscopy. The dependence of L_{rel} and $\ln T_2$ on $\ln[M_\Sigma]$ (Fig. 8) was used to estimate sizes of the primary and secondary nuclei.

The renormalized curves presented in Fig. 7 are very similar to one another, in accordance with [26, 33]. They have large lag periods, which rules out the linear growth scenario.

Table 2 summarizes results obtained for amyloids composed of proteins and peptides examined in [10] and of those experimentally studied by us [21]. Only one of them (apolipoprotein C-II [34]) exhibits linear growth ($L_{rel}<0.1$, the primary nucleus size $n^* \approx 4$–5). Growth with bifurcation is found for one peptide (Aβ42 [17]) and one protein (Titin domain TI I27 [35]); for both of them, the secondary nucleus size $n_2 \approx 2$ and the primary nucleus size $n^* \approx 3$. In all the other cases, kinetic data taken alone are consistent with exponential growth with either fragmentation or bifurcation, with $n_2=0$; $n^* \approx 1$–2 for all of them. Examination of EM images of

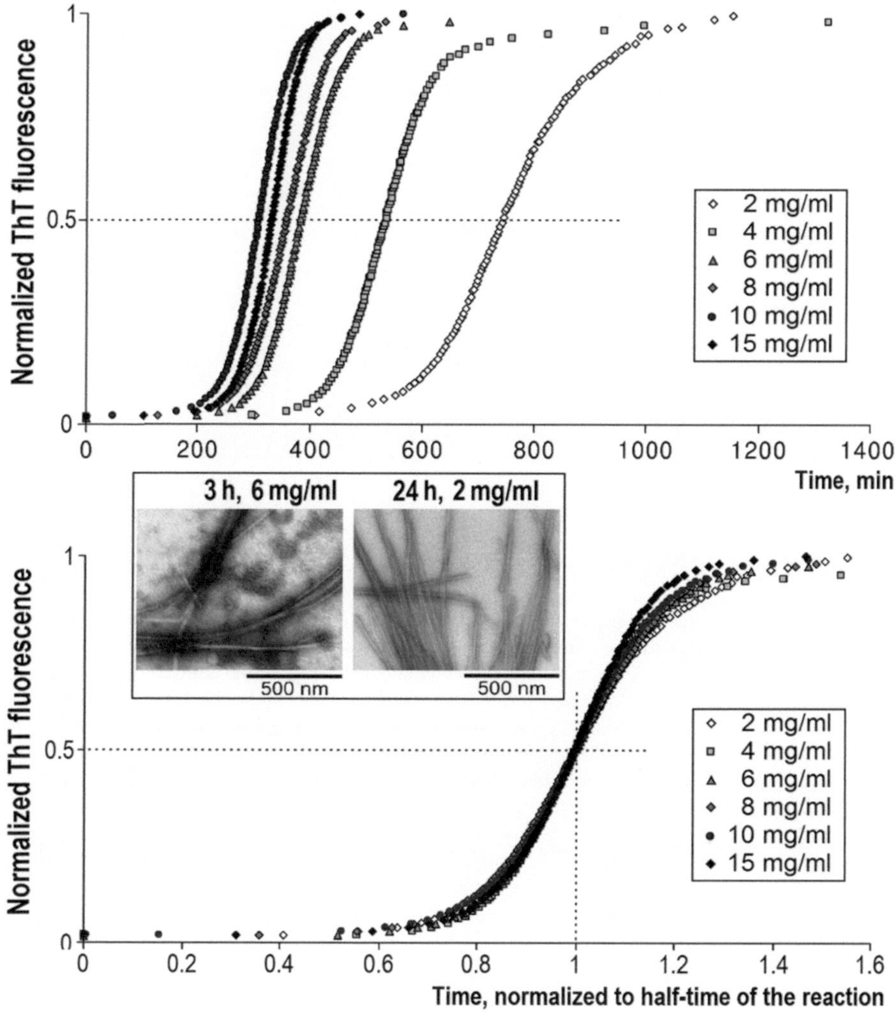

Fig. 7 ThT fluorescence and EM data for human insulin. *Upper panel*: ThT fluorescence is presented as a function of incubation time at different insulin concentrations. To obtain a curve for each protein concentration, 4–5 experiments have been made (each of them normalized to the time of reaching 50 % of the maximal ThT fluorescence). *Lower panel*: Lines shown in the upper panel are represented as functions of the incubation time normalized to the time of reaching 50 % of the maximal ThT fluorescence for the given insulin concentration. *Inset*: Traces of bifurcation at EM images of insulin after 6 h incubation at the 6 mg/mL concentration and after 24 h incubation at the 2 mg/mL concentration. Small amorphous aggregates that appear at early stages of amyloidogenesis disappear at later stages. Adapted from [21]

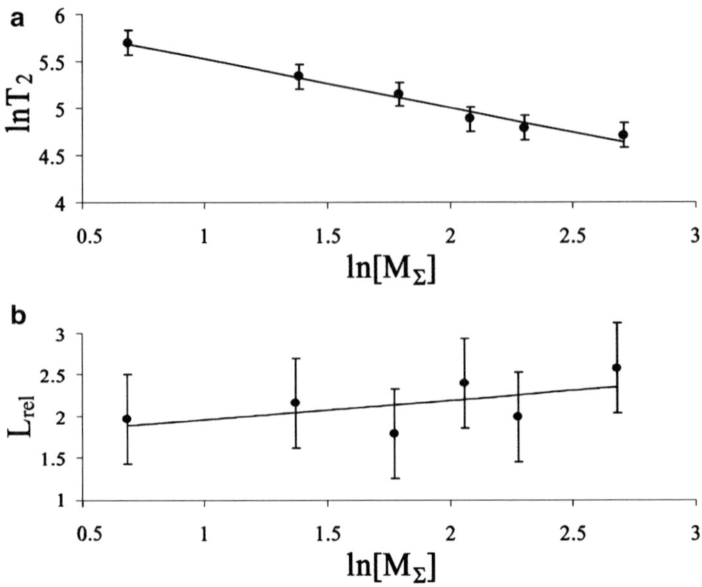

Fig. 8 The approximation line calculated for insulin amyloid formation (with experimental data from Fig. 7) by the least-squares method plotted in coordinates (**a**) {$\ln T_2$, $\ln[M_\Sigma]$} and (**b**) {L_{rel}, $\ln[M_\Sigma]$}; $[M_\Sigma]$ is in mg/mL, T_2 in minutes. The tangent coefficients of the approximation lines are -0.52 ± 0.13 for $d(\ln T_2)/d(\ln[M_\Sigma])$ and 0.23 ± 0.52 for $d(L_{rel})/d(\ln[M_\Sigma])$; the errors are computed with the Student's coefficients corresponding to 0.95 confidence level. The value $d(\ln T_2)/d(\ln[M_\Sigma]) \approx -0.5$ is inconsistent with the "growth from the surface" model but consistent with "linear growth" (which, though, should have $L_{rel} < 0.2$, which is inconsistent with L_{rel}), "fragmentation" (which, though, is inconsistent with EM images shown in *Inset* of Fig. 7 and Figs. 1, 2 and 3 of [21]), and "bifurcation" (consistent with those EM images), with the secondary nucleus size $n_2 = 0$ calculated as $-1 - d(\ln T_2)/d(\ln[M_\Sigma])$; the primary nucleus size $n^* = 1 + n_2 - d(L_{rel})/d(\ln[M_\Sigma]) = 1$ (see Tables 1 and 2). Adapted from [21]

growing insulin fibrils shows [21] that they accord with a kind of the "bifurcation plus lateral growth [15]" scenario. The "growth from the surface" scenario has not been observed in the small set of peptides and proteins listed in Table 2.

Conclusions

We have shown that kinetic curves can give an insight into various mechanisms of amyloid aggregate formation. Small lag-periods ($L_{rel} < 0.2$) are typical of the linear growth, while for all exponential growth scenarios the values of L_{rel} are larger, and usually significantly larger. The dependences of $\ln T_2$ and L_{rel} on $\ln[M_\Sigma]$ (Table 1) help to distinguish among different exponential growth scenarios (though an

Table 2 The range of L_{rel} and $\ln T_2$ values, the sizes of the primary (n^*) and secondary (n_2) nuclei and the amyloid growth scenarios obtained from experimental kinetic data alone

Amyloid-forming protein or peptide and a ref. to its experimental study	$[M_\Sigma]$, mg/mL (min–max)[a]	$\ln(T_2/\text{min})$ at various $[M_\Sigma]$ (min–max)	L_{rel} at various $[M_\Sigma]$ (min–max)	$n^* \pm \Delta n^*$, primary nucleus[b]	$n_2 \pm \Delta n_2$, secondary nucleus[b]	Model that follows from the L_{rel} and $\ln T_2$ on $[M_\Sigma]$ dependencies
Human insulin [21] (37 °C)[c]	2.0–15	4.71–5.70	1.78–2.48	0.81 ± 0.54	0.04 ± 0.13	Exponential[d]
Human insulin [26] (45 °C)[e]	2.5–20	3.60–4.18	5.17–5.56	1.13 ± 0.19	−0.48 ± 0.16	Exponential[d]
LysPro mutant [21] of human insulin (37 °C)[c]	2.0–15	5.78–4.81	2.04–4.38	0.51 ± 2.05	−0.20 ± 0.51	Exponential[d]
Aβ42 [17]	0.005–0.02	3.58–5.53	0.53–0.67	2.64 ± 0.11	1.72 ± 0.05	Bifurcation (with $n_2 \approx 2$)
Yeast prion Sup35 [22]	0.03–0.34	4.62–5.52	0.29–0.70	1.09 ± 0.20	−0.51 ± 0.45	Exponential[d]
Yeast Prion Ure2p [27]	0.36–1.53	5.94–5.37	0.76–1.02	0.96 ± 1.52	−0.23 ± 1.40	Exponential[d]
Murine FBP28 WW domain [28]	0.25–2.50	4.76–6.02	1.56–2.10	1.21 ± 1.27	0.05 ± 1.02	Exponential[d]
β2-micro-globulin[9][f]	0.10–2.86	4.88–6.38	1.48–3.86	1.58 ± 0.58	−0.06 ± 0.18	Exponential[d]
Apolipo-protein C-II [34]	0.20–0.50	6.87–8.91	0.06–0.10	4.44 ± 0.38	–	Linear
Titin domain TI I27 [35]	1.0–4.0	1.84–3.76	0.14–0.34	2.86 ± 0.30	2.04 ± 0.29	Bifurcation (with $n_2 \approx 2$)

[a]We used only those amyloid formation kinetics where, according to remarks given in the original papers, characteristic times can be reliably estimated

[b]Errors are computed using the Student's coefficient corresponding to 0.95 confidence level

[c]In 20 % acetic acid, 0.14 M NaCl (pH 2.0); non-aggregated insulin is known to be monomeric under these conditions [32]

[d]Fragmentation or bifurcation with $n_2 = 0$

[e]In 20 % acetic acid, 0.5 M NaCl (pH 1.8)

[f]Numerical data for β2-microglobulin amyloidogenesis kinetics were kindly provided by S.E. Radford

additional direct observation of fibril shapes is needed in some cases) and to establish the sizes of the primary and secondary folding nuclei of fibrils.

It should be added that a particular scenario used by the given protein to form an amyloid may depend not only on the protein, solvent conditions and agitation

[36], but also on protein concentration. Indeed, the exponential scenarios take place when the primary fibril nucleation by separate monomers is very slow, as compared to the secondary fibril initiation (e.g., by fragmentation, which works when $\lambda_+ >> k_+ [M_\Sigma]^{n^*-1}$, see (13)). Thus, if the total protein concentration $[M_\Sigma]$ is very low, a kind of the exponential growth is most probable, even if the linear growth is observed at a high $[M_\Sigma]$.

A typical time of amyloid formation in experiment lasts for hours (Fig. 7), sometimes days [37]. However, it is remarkable that a moderate increase (by ≈10 kcal/mol only) in the energy of fibril initiation (i.e., in parameters k_+ and λ_+), as well as a low concentration of the amyloidogenic protein (especially when the primary and secondary folding nuclei of fibrils are large), can convert these hours or days in many years or many decades (see (2), (13), (15)). This may explain an enormous incubation time of the amyloid-connected diseases [6, 14].

In closing, it may be said that the above considered models correspond to "pure" cases of amyloid formation by different mechanisms, while a combination of several mechanisms is quite possible and can be observed sometimes in experiments [18].

Acknowledgments We are grateful to S. E. Radford who kindly provided us with numerical data for β2-microglobulin amyloidogenesis kinetics. We thank E. V. Serebrova for assistance in manuscript preparation. This study was supported in part by the Russian Science Foundation Grants 14-24-00157, 14-14-00536.

References

1. Oosawa F, Asakura S, Hotta K, Imai N, Ooi T (1959) G-F transformation of actin as a fibrous condensation. J Polym Sci 37:323–336
2. Frieden C, Goddette DW (1983) Polymerization of actin and actin-like systems: evaluation of the time course of polymerization in relation to the mechanism. Biochemistry 22:5836–5843
3. Goldstein RF, Stryer L (1986) Cooperative polymerization reactions. Analytical approximations, numerical examples, and experimental strategy. Biophys J 50:583–599
4. Hofrichter J, Ross PD, Eaton WA (1974) Kinetics and mechanism of deoxyhemoglobin S gelation: a new approach to understanding sickle cell disease. Proc Natl Acad Sci U S A 71:4864–4868
5. Ferrone FA, Hofrichter J, Sunshine HR, Eaton WA (1980) Kinetic studies on photolysis-induced gelation of sickle cell hemoglobin suggest a new mechanism. Biophys J 32:361–380
6. Buxbaum JN, Linke RP (2012) A molecular history of the amyloidoses. J Mol Biol 421:142–159
7. Cali I, Castellani R, Yuan J, Al-Shekhlee A, Cohen ML, Xiao X, Moleres FJ, Parchi P, Zou WQ, Gambetti P (2006) Classification of sporadic Creutzfeldt-Jakob disease Revisited. Brain 129:2266–2277
8. Serio TR, Cashikar AG, Kowal AS, Sawicki GJ, Moslehi JJ, Serpell L, Arnsdorf MF, Lindquist SL (2000) Nucleated conformational conversion and the replication of conformational information by a prion determinant. Science 289:1317–1321

9. Xue WF, Homans SW, Radford SE (2008) Systematic analysis of nucleation-dependent polymerization reveals new insights into the mechanism of amyloid self-assembly. Proc Natl Acad Sci U S A 105:8926–8931
10. Knowles TP, Waudby CA, Devlin GL, Cohen SI, Aguzzi A, Vendruscolo M, Terentjev EM, Welland ME, Dobson CM (2009) An analytical solution to the kinetics of breakable filament assembly. Science 326:1533–1537
11. Cohen SI, Vendruscolo M, Welland ME, Dobson CM, Terentjev EM, Knowles TP (2011) Nucleated polymerization with secondary pathways. I. Time evolution of the principal moments. J Chem Phys 135:065105
12. Morris AM, Watzky MA, Finke RG (2009) Protein aggregation kinetics, mechanism, and curve-fitting: a review of the literature. Biochim Biophys Acta 1794:375–397
13. Giehm L, Otzen DE (2010) Strategies to increase the reproducibility of protein fibrillization in plate reader assays. Anal Biochem 400:270–281
14. Masel J, Jansen VA, Nowak MA (1999) Quantifying the kinetic parameters of prion replication. Biophys Chem 77:139–152
15. Loksztejn A, Dzwolak W (2010) Vortex-induced formation of insulin amyloid superstructures probed by time-lapse atomic force microscopy and circular dichroism spectroscopy. J Mol Biol 395:643–655
16. Andersen CB, Yagi H, Manno M, Martorana V, Ban T, Christiansen G, Otzen DE, Goto Y, Rischel C (2009) Branching in amyloid fibril growth. Biophys J 96:1529–1536
17. Cohen SI, Linse S, Luheshi LM, Hellstrand E, White DA, Rajah L, Otzen DE, Vendruscolo M, Dobson CM, Knowles TP (2013) Proliferation of amyloid-β42 aggregates occurs through a secondary nucleation mechanism. Proc Natl Acad Sci U S A 110:9758–9763
18. Jeong JS, Ansaloni A, Mezzenga R, Lashuel HA, Dietler G (2013) Novel mechanistic insight into the molecular basis of amyloid polymorphism and secondary nucleation during amyloid formation. J Mol Biol 425:1765–1781
19. Cho KR, Huang Y, Yu S, Yin S, Plomp M, Qiu SR, Lakshminarayanan R, Moradian-Oldak J, Sy M-S, De Yoreo JJ (2011) A multistage pathway for human prion protein aggregation in vitro: from multimeric seeds to β oligomers and nonfibrillar structures. J Am Chem Soc 133:8586–8593
20. Dovidchenko NV, Finkelstein AV, Galzitskaya OV (2014) How to determine the size of folding nuclei of protofibrils from the concentration dependence of the rate and lag-time of aggregation. I. Modeling the amyloid photofibril formation. J Phys Chem B 118:1189–1197
21. Selivanova OM, Suvorina MY, Dovidchenko NV, Eliseeva IA, Surin AK, Finkelstein AV, Schmatchenko VV, Galzitskaya OV (2014) How to determine the size of folding nuclei of protofibrils from the concentration dependence of the rate and lag-time of aggregation. II. Experimental application for insulin and lys-pro insulin: aggregation morphology, kinetics and sizes of nuclei. J Phys Chem B 118:1198–1206
22. Collins SR, Douglass A, Vale RD, Weissman JS (2004) Mechanism of prion propagation: amyloid growth occurs by monomer addition. PLoS Biol 2:e321
23. Eyring H (1935) The activated complex in chemical reactions. J Chem Phys 3:107–115
24. Moore JW, Pearson RG (1981) Kinetics and mechanism. Wiley, Chichester
25. Finkelstein AV (2014) Time to overcome the high, long and bumpy free-energy barrier in a multi-stage process: the generalized steady-state approach. J Phys Chem B 119:158–163
26. Fodera V, Librizzi F, Groenning M, van de Weert M, Leone M (2008) Secondary nucleation and accessible surface in insulin amyloid fibril formation. J Phys Chem B 112:3853–3858
27. Zhu L, Zhang XJ, Wang LY, Zhou JM, Perrett S (2003) Relationship between stability of folding intermediates and amyloid formation for the yeast prion Ure2p: a quantitative analysis of the effects of pH and buffer system. J Mol Biol 328:235–254
28. Ferguson N, Berriman J, Petrovich M, Sharpe TD, Finch JT, Fersht AR (2003) Rapid amyloid fiber formation from the fast-folding WW domain Fbp28. Proc Natl Acad Sci U S A 100:9814–9819

29. Bishop MF, Ferrone FA (1984) Kinetics of nucleation-controlled polymerization. A perturbation treatment for use with a secondary pathway. Biophys J 46:631–644
30. Wegner A (1982) Spontaneous fragmentation of actin filaments in physiological conditions. Nature 296:266–267
31. Castilla J, Saá P, Hetz C, Soto C (2005) In vitro generation of infectious scrapie prions. Cell 121:195–206
32. Nielsen L, Frokjaer S, Brange J, Uversky VN, Fink AL (2001) Probing the mechanism of insulin fibril formation with insulin mutants. Biochemistry 40:8397–8409
33. Fändrich M (2007) Absolute correlation between lag time and growth rate in the spontaneous formation of several amyloid-like aggregates and fibrils. J Mol Biol 365:1266–1270
34. Binger KJ, Pham CLL, Wilson LM, Bailey MF, Lawrence LJ, Schuck P, Howlett GJ (2008) Apolipoprotein C-II amyloid fibrils assemble via a reversible pathway that includes fibril breaking and rejoining. J Mol Biol 376:1116–1129
35. Wright CF, Teichmann SA, Clarke J, Dobson CM (2005) The importance of sequence diversity in the aggregation and evolution of proteins. Nature 438:878–881
36. Yoshimura Y, Lin Y, Yagi H, Lee Y-H, Kitayama H, Sakurai K, So M, Ogi H, Naiki H, Goto Y (2012) Distinguishing crystal-like amyloid fibrils and glass-like amorphous aggregates from their kinetics of formation. Proc Natl Acad Sci U S A 109:14446–14451
37. Zurdo J, Guijarro JI, Jiménez JL, Saibil HR, Dobson CM (2001) Dependence on solution conditions of aggregation and amyloid formation by an SH3 domain. J Mol Biol 311:325–340

Modulation of Cholera Toxin Structure/Function by Hsp90

Michael Taylor, Christopher B.T. Britt, Jennifer Fundora, and Ken Teter

Abstract Cholera toxin (CT) is released into the extracellular environment, but the catalytic CTA1 subunit attacks its G protein target within the cytosol of an intoxicated cell. To access the cytosol, CT moves by vesicle carriers from the cell surface to the endoplasmic reticulum (ER). CTA1 then dissociates from the rest of the toxin and shifts to a disordered conformation that facilitates its passage into the cytosol through a pore in the ER membrane. We have found that CTA1 export to the cytosol requires the host cytosolic chaperone Hsp90. Loss of Hsp90 function trapped CTA1 in the ER, and Hsp90 was sufficient for in vitro export of CTA1 from the ER. Structural studies demonstrated Hsp90 will refold disordered CTA1. ATP hydrolysis by Hsp90 was required for both CTA1 refolding and CTA1 extraction from the ER, which suggests a ratchet mechanism for the chaperone-driven movement of CTA1 to the cytosol: the refolding of CTA1 as it emerges at the cytosolic face of the ER membrane would prevent the toxin from sliding back into the translocon pore and would thereby ensure the unidirectional movement of CTA1 from the ER to the cytosol. Hsp90 bound to the N-terminus of CTA1 and did not release CTA1 after refolding the toxin. The continued association of Hsp90 with CTA1 allowed the toxin to maintain an active conformation at 37 °C. Hsp90 thus plays two key roles CT intoxication: it couples toxin refolding with toxin extraction from the ER, and it maintains the cytosolic toxin in a functional conformation.

Keywords AB toxins • Chaperones • Cholera toxin • Endoplasmic reticulum • Endoplasmic reticulum-associated degradation • Exotoxin A • Fourier transform infrared spectroscopy • Host-pathogen interactions • Hsp90 • Shiga toxin • Surface plasmon resonance • Toxin activity • Toxin structure

---

M. Taylor • C.B.T. Britt • K. Teter (✉)
Burnett School of Biomedical Sciences, College of Medicine,
University of Central Florida, Orlando, FL 32826, USA
e-mail: kteter@mail.ucf.edu

J. Fundora
Burnett School of Biomedical Sciences, College of Medicine,
University of Central Florida, Orlando, FL 32826, USA

Johns Hopkins Children's Center, Baltimore, MD 21287, USA

L. Olivares-Quiroz et al. (eds.), *Physical Biology of Proteins and Peptides*,
DOI 10.1007/978-3-319-21687-4_4

Structure and Function of Cholera Toxin

Cholera toxin (CT) is a multimeric AB-type protein toxin that contains a catalytic A1 subunit, an A2 linker, and a cell-binding B homopentamer (Fig. 1) [1]. The A1 and A2 subunits are synthesized as a single polypeptide that is proteolytically nicked to generate a disulfide-linked A1/A2 heterodimer. CTA1 is an ADP-ribosyltransferase, while CTA2 extends into the central pore of the ring-like CTB pentamer and thereby anchors CTA1 to the cell-binding moiety of the toxin. The CT holotoxin is assembled in the bacterial periplasm and released into the extracellular environment where it contacts its GM1 ganglioside receptor on the host plasma membrane. Although the entire CT holotoxin enters the host cell by receptor-mediated endocytosis, only the CTA1 subunit reaches the cytosol.

The cytosolic target for CTA1 is Gsα, a signaling protein that stimulates adenylate cyclase to produce cAMP [2, 3]. Gsα is a component of the heterotrimeric G protein, and it oscillates between active and inactive states based on the status of its bound guanine nucleotide. When Gsα is ADP-ribosylated by CTA1 at specific arginine residue, it cannot hydrolyze its bound GTP and is consequently locked in an active conformation. This results in continual stimulation of adenylate cyclase, persistently high levels of intracellular cAMP, and ongoing activation of the signaling pathways responsible for chloride efflux and sodium malabsorption in the small intestine [4]. The osmotic movement of water which follows salt accumulation in the gut then generates the profuse, potentially life-threatening diarrhea of cholera [5].

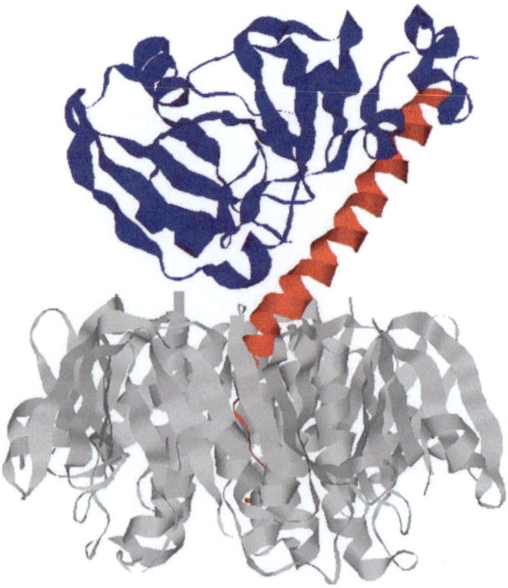

Fig. 1 CT structure. The 21 kDa CTA1 subunit (*blue*) is anchored to the 5 kDa CTA2 linker (*red*) by numerous non-covalent contacts and a single disulfide bond near the C-terminus of CTA1 and N-terminus of CTA2. CTA2 extends into the central pore of the ring-like CTB homopentamer (*grey*) and thus maintains extensive non-covalent contacts with CTB as well. The image was derived from PDB 1S5F

CT Transport from the Cell Surface to the ER

To activate the cAMP-dependent signaling pathways that produce a diarrheatic effect, CTA1 must reach Gsα in the cytosol of an intoxicated cell. This process begins when the CT holotoxin adheres to GM1 gangliosides on the surface of a target cell and is internalized by an endocytic mechanism involving lipid rafts [6]. Although the majority of internalized CT is routed from the endosomes to the lysosomes for degradation, a minor pool (~10 %) of cell-associated toxin moves by vesicle carriers from the endosomes to the Golgi apparatus en route to the endoplasmic reticulum (ER).

The CTA1/CTA2 disulfide bond is reduced in the ER, but reduction alone is insufficient for holotoxin disassembly: numerous non-covalent contacts between CTA1 and the CTA2/CTB_5 complex are sufficient to preserve a stable, intact holotoxin even after reduction of the A1/A2 disulfide bond [7]. Disassembly of the reduced holotoxin instead results from an interaction with protein disulfide isomerase (PDI), an ER-localized protein with linked but independent oxidoreductase and chaperone functions [8]. PDI unfolds upon contact with CTA1, and we have proposed the expanded hydrodynamic radius of unfolded PDI acts as a lever to dislodge reduced CTA1 from its non-covalent assembly with the rest of the toxin [9]. In support of this model, PDI-mediated disassembly of the reduced holotoxin did not occur when PDI was locked in a folded conformation by treatment with the "zero-length" intramolecular cross-linker EDC or when its substrate-induced unfolding was blocked by ribostamycin, an aminoglycoside antibiotic that specifically blocks the chaperone function of PDI. Bacitracin, a peptide antibiotic that blocks the reductive activity of PDI, did not inhibit the substrate-induced unfolding of PDI and did not disrupt PDI-mediated disassembly of the reduced holotoxin. Both the substrate-induced unfolding of PDI and PDI-mediated toxin disassembly were thus linked to the chaperone function of PDI. Furthermore, the chaperone activity of PDI is essential for CT intoxication: PDI-deficient cells and cells incubated with ribostamycin are resistant to the toxin [9, 10]. Two other ER-localized oxidoreductases (ERp57 and ERp72) can bind to CT but do not unfold upon contact with CTA1 and do not displace reduced CTA1 from CTA2/CTB_5 [9]. PDI thus plays a unique, critical role in the CT intoxication process.

CTA1 Translocation from the ER to the Cytosol

CTA1 maintains a folded conformation when incorporated into the CT holotoxin, but the free CTA1 subunit is a thermally unstable protein [11]. Holotoxin disassembly at physiological temperature thus leads to spontaneous unfolding of the released A1 subunit. The dissociated, disordered CTA1 polypeptide is consequently recognized by the quality control mechanism of ER-associated degradation (ERAD).

This system identifies misfolded or misassembled proteins in the ER and exports them to the cytosol for ubiquitin-dependent proteasomal degradation [12, 13]. ERAD chaperones such as BiP and ERdj3 maintain CTA1 in a soluble state before its export to the cytosol through Hrd1 and/or Sec61 translocon pores [14]. CTA1 avoids the rapid proteolysis which usually accompanies ERAD-mediated translocation to the cytosol because the arginine-over-lysine bias in its amino acid sequence shields it from modification with the polyubiquitin chains that are appended to lysine residues [15]. CTA1 is eventually cleared from the cytosol by a ubiquitin-independent proteasomal mechanism [11], but its evasion of efficient turnover by the ubiquitin-proteasome system allows it to remain in the cytosol long enough to elicit a cytopathic effect.

Most ERAD substrates are exported from the ER in a process involving the AAA ATPase p97 and the associated Ufd1/Npl4 heterodimer [16]. A circular, p97 hexamer docks at the cytosolic face of the translocon pore and uses the energy from ATP hydrolysis to extract proteins from the ER. Ubiquitinated ERAD substrates are recognized during the translocation event by Ufd1/Npl4, which helps recruit p97 to the membrane. The combined action of p97 and Ufd1/Npl4 thus promotes ERAD-mediated translocation to the cytosol. However, neither p97 nor Ufd1/Npl4 interact with CTA1 in the expected manner. Loss of p97 function due to RNAi or the expression of a dominant negative construct only produced a minimal inhibitory effect on CT intoxication, whereas cells lacking Ufd1 or Npl4 due to RNAi were more sensitive to CT than the control cells [17, 18]. The latter result suggests Ufd1/Npl4 acts as a negative regulator of CTA1 translocation. Since CTA1 export to the cytosol does not involve direct ubiquitination of CTA1 [15] or the ubiquitin machinery at all [11], the role of Ufd1/Npl4 in CTA1 translocation appears to be separate from its normal p97- and ubiquitin-linked functions.

Since the p97 complex is not responsible for the ER-to-cytosol export of CTA1, we hypothesized Hsp90 was involved with the translocation event. Hsp90 acts as an ATP-dependent dimer to stabilize and/or fold client proteins in the eukaryotic cytosol [19]. Many Hsp90 client proteins function in oncogenesis and are highly expressed in cancer cells, so Hsp90 has been targeted for chemotherapy with geldanamycin (GA) and other drugs that disrupt ATP binding to the chaperone [20]. Most of these therapeutic agents also target GRP94, an ER-localized homolog of Hsp90. However, one drug affects GRP94 but not Hsp90: N-ethylcarboxamidoadenosine (NECA). Parallel experiments performed with GA and NECA can therefore distinguish between Hsp90-dependent and GRP94-dependent cellular effects.

In addition to cancer, Hsp90 has also been linked to infectious disease: it is required for certain AB toxins to enter the cytosol from acidified endosomes [21, 22]. Hsp90 can also assist the ERAD processing of misfolded transmembrane proteins [23, 24], and it was shown to refold an exogenously added protein that entered the cytosol in a disordered conformation after moving from the cell surface to the ER by vesicular transport [25]. These collective observations suggested Hsp90 could play a role in the ER-to-cytosol export of CTA1.

The Functional Role of Hsp90 in CTA1 Translocation

We found cells lacking Hsp90 due to RNAi were highly resistant to CT. GA-treated cells were highly resistant to CT as well, but cells treated with NECA maintained sensitivity to the toxin [26]. GRP94 was thus dispensable for CT intoxication. GA also inhibited CT activity in the ileal loop model of intoxication: co-injection of GA with CT into surgically sealed sections of rat small intestine greatly reduced the level of fluid accumulation in comparison to intestinal sections injected with CT alone. Thus, the GA derivatives that are in clinical trials as anti-cancer agents could potentially be repurposed as anti-toxin agents for cholera treatment.

Additional studies demonstrated the loss of Hsp90 function protects cells from CT by blocking CTA1 access to the cytosol [26]. For this work, a toxin detection system based on the technique of surface plasmon resonance (SPR) was used to monitor CTA1 accumulation in the cytosol of untreated or drug-treated cells [27]. Selective permeabilization of the plasma membrane with digitonin, followed by a low-speed centrifugal spin, produced distinct organelle and cytosolic fractions from intoxicated cells (Fig. 2a). Cytosolic fractions were then perfused over an SPR slide

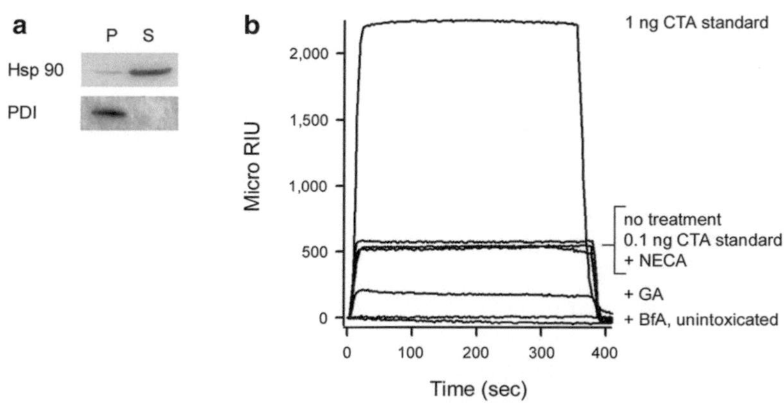

Fig. 2 Hsp90 but not GRP94 is required for CTA1 translocation to the cytosol. HeLa cells were pulse-labeled at 4 °C for 30 min with 1 μg/mL of CT. The cells were then chased for 2 h at 37 °C in toxin-free medium containing no additions, 0.1 μM GA, 0.1 μM NECA, or 5 μg/mL of BfA. Permeabilization of the plasma membrane with digitonin was used to partition cell extracts into separate organelle (pellet; P) and cytosolic (supernatant; S) fractions. (**a**) Both fractions from untreated cells were probed by Western blot to establish the distributions of cytosolic marker Hsp90 and lumenal ER marker PDI. (**b**) An SPR sensor coated with an anti-CTA antibody was used to detect the cytosolic pool of CTA1 from untreated or drug-treated cells. CTA standards were perfused over the sensor slide as positive controls, and the cytosolic fraction from unintoxicated cells was perfused over the sensor as a negative control. At the end of each perfusion, bound sample was stripped from the sensor slide. This research was originally published in the *Journal of Biological Chemistry*. Taylor M, Navarro-Garcia F, Huerta J, Burress H, Massey S, Ireton K, and Teter K. Hsp90 is Required for Transfer of the Cholera Toxin A1 Subunit from the Endoplasmic Reticulum to the Cytosol. *J Biol Chem*. 2010; 285(41):31261–31267. © the American Society for Biochemistry and Molecular Biology

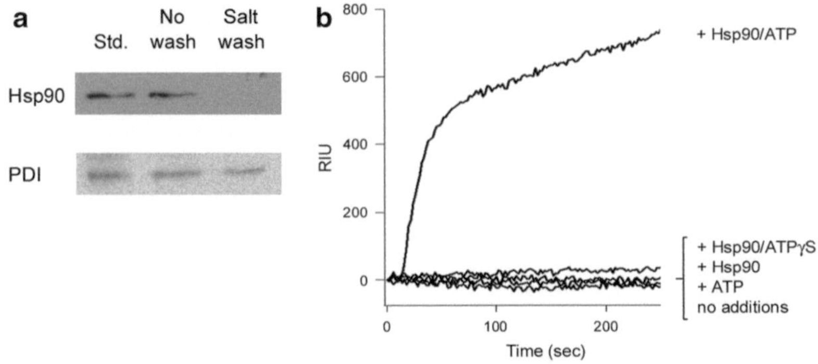

Fig. 3 Hsp90/ATP is sufficient for CTA1 extraction from the ER. CHO cells incubated with 100 ng/mL of CT for 2 h at 37 °C were treated with digitonin to selectively permeabilize the plasma membrane. Following centrifugation, the intact membrane pellet was collected for further study. (**a**) Western blot analysis was used to detect Hsp90 and PDI, a lumenal ER protein, in association with unwashed or salt-washed membrane pellets. Hsp90 and PDI protein standards (Std.) were used as controls. (**b**) Purified Hsp90, ATP, Hsp90/ATP, or Hsp90/ATPγS was added to the salt-washed membrane pellet for 1 h at 37 °C. To detect the exported pool of CTA1, supernatant samples obtained after centrifugation were perfused over an SPR sensor coated with an anti-CTA antibody. This research was originally published in the *Journal of Biological Chemistry*. Burress H, Taylor M, Banerjee T, Tatulian SA, and Teter K. Co- and Post-Translocation Roles for Hsp90 in Cholera Intoxication. *J Biol Chem*. 2014; 289(48):33644–33654. © the American Society for Biochemistry and Molecular Biology

coated with an anti-CTA antibody (Fig. 2b). CTA1 was detected in the cytosol of untreated cells but was absent from the cytosol of cells treated with brefeldin A (BfA), a drug that prevents CTA1 delivery to the cytosol by blocking CT transport from the endosomes to the ER translocation site [28]. This control demonstrated the cytosolic pool of CTA1 from untreated cells did not result from rupture of toxin-containing organelles during the fractionation procedure. Likewise, the lack of signal from unintoxicated cells demonstrated our SPR sensor specifically recognized CTA1 and not a component of the host cytosol. Equivalent levels of CTA1 could be detected in the cytosol of untreated or NECA-treated cells. In contrast, 2.4-fold less CTA1 was detected in the cytosol of GA-treated cells. A similar 3.2-fold drop in cytosolic CTA1 was recorded for cells lacking Hsp90 due to RNAi, and additional control experiments demonstrated that toxin transport to the ER was not affected by the loss of Hsp90 function [26]. These collective results, which were consistent with our toxicity data, indicated Hsp90 but not GRP94 is required for CTA1 passage into the cytosol. This provided the first demonstration for the role of Hsp90 in the ERAD-mediated extraction of a soluble protein from the ER.

An in vitro translocation assay further demonstrated that Hsp90 is sufficient for CTA1 translocation to the cytosol [29]. For this work, CHO cells were incubated in CT-containing medium for 3 h. Selective permeabilization of the plasma membrane with digitonin was again used to generate distinct organelle and cytosolic fractions, but in this case we collected the endomembrane-associated toxin in the pellet/organelle fraction. Washes with high salt concentrations were used to strip Hsp90 from the cytosolic face of the pelleted ER membranes (Fig. 3a), and purified Hsp90 was then added to the pellet fraction for 1 h at 37 °C. Another low speed centrifugal spin separated the ER from the supernatant, and a SPR sensor coated with an anti-CTA antibody was used to detect any CTA1 exported to the supernatant during the previous hour at 37 °C. CTA1 did not appear in the supernatant when the toxin-containing membranes were incubated with buffer alone, buffer containing ATP, or buffer containing Hsp90 in the absence of ATP. In contrast, CTA1 translocation occurred in the presence of Hsp90/ATP (Fig. 3b). This demonstrated Hsp90/ATP is sufficient for CTA1 export from the ER to the cytosol. Since our previous in vivo work established the essential role of Hsp90 in CTA1 translocation to the cytosol, our collective studies have shown that Hsp90 is necessary and sufficient for the ER-to-cytosol export of CTA1.

The Structural Role of Hsp90 in CTA1 Translocation

Biophysical studies on the structural impact of Hsp90 binding to CTA1 have provided a possible mechanistic basis for the role of Hsp90 in CTA1 translocation [29]. Using the technique of isotope-edited Fourier transform infrared spectroscopy, we demonstrated Hsp90 induces a gain-of-structure in the disordered CTA1 polypeptide: the 37 °C conformation of CTA1 shifted from 17 % α-helix, 15 % β-sheet, and 54 % irregular (i.e., undefined) structure to 28 % α-helix, 33 % β-sheet, and 28 % irregular structure upon the addition of Hsp90/ATP. For comparison, the folded conformation of CTA1 at 10 °C contained 34 % α-helix, 42 % β-sheet, and 17 % irregular structure. The partial refolding of CTA1 by Hsp90 did not occur in the absence of ATP or in the presence of non-hydrolyzable ATPγS. Control experiments ensured Hsp90/ATPγS could bind to CTA1, so ATP hydrolysis was critical for the Hsp90-mediated refolding of CTA1.

Hsp90 Couples CTA1 Refolding with CTA1 Extraction from the ER

Our biophysical studies demonstrated ATP hydrolysis by Hsp90 is required for the refolding of disordered CTA1. Our in vitro translocation assay further demonstrated that ATP hydrolysis by Hsp90 is required for the ER-to-cytosol export of CTA1: in contrast to Hsp90/ATP, Hsp90/ATPγS could not extract CTA1 from the ER (Fig. 3b). Based on these two linked observations, we proposed a model for toxin translocation in which Hsp90 couples CTA1 refolding with CTA1 extraction from the ER. The Hsp90-mediated refolding of CTA1 as the toxin emerges at the cytosolic face of the ER membrane would prevent the refolded toxin from sliding back into the narrow diameter of the translocon pore, thus ensuring the unidirectional ER-to-cytosol movement of CTA1. Loss of Hsp90 function would accordingly trap CTA1 in the ER, as we have observed. This model is conceptually similar to the

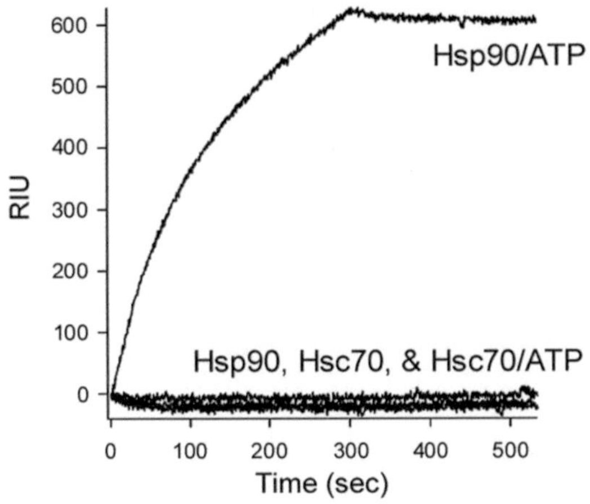

Fig. 4 Hsp90 binds to the N-terminus of CTA1. Hsp90 or Hsc70 in either the absence or presence of ATP was perfused over a SPR slide coated with a peptide corresponding to the 16 N-terminal amino acids of the mature CTA1 polypeptide (i.e., after removal of the CTA leader sequence). Ligand was removed from the perfusion buffer after 300 s. Experiments were performed with a Reichert SR7000 SPR Refractometer following the general procedure described in Pande et al. [11]

"Brownian ratchet" mechanism by which ER chaperones and folding enzymes direct co-translational protein import into the ER [30].

As a logical extension of our model, Hsp90 would be expected to bind the toxin as soon as the N- or C-terminus of CTA1 appears at the cytosolic face of the ER membrane. Preliminary experiments have indeed found that Hsp90 recognizes the N-terminal region of CTA1. Hsp90 was captured by a SPR sensor coated with a peptide corresponding to the N-terminal 16 amino acid residues of CTA1 (Fig. 4), but it did not bind to a SPR sensor coated with a peptide corresponding to the C-terminal 16 amino acid residues of CTA1 (not shown). Binding to the N-terminal CTA1 peptide was a specific event, as it did not occur in the absence of ATP. Hsc70, another chaperone of the host cytosol, did not bind to the $CTA1_{1-16}$ peptide in either the presence or absence of ATP (Fig. 4). This preliminary experiment indicates Hsp90 is recruited to the CTA1 polypeptide as soon as the toxin N-terminal domain emerges from the translocation pore. The data further suggest CTA1 translocation proceeds in an N- to C-terminal direction, which is consistent with the translocation-competent state of a CTA1 deletion construct lacking the C-terminal 23 amino acid residues of the toxin [31]. The current experimental evidence thus supports a translocation model in which CTA1 extraction from the ER results from the Hsp90-driven refolding of CTA1 that occurs after chaperone binding to the N-terminal region of the toxin.

Post-translocation Association Between Hsp90 and CTA1

SPR binding assays documented a high affinity interaction between CTA1 and Hsp90 (K_D=7 nM). These studies further noted (1) a dimer of Hsp90 binds to CTA1; (2) Hsp90 cannot bind to CTA1 in the absence of ATP or ATPγS; (3) Hsp90 recognizes unfolded but not folded conformations of CTA1; and (4) Hsp90 does not dissociate from CTA1 after ATP hydrolysis and refolding of the toxin [26, 29]. The first three observations were consistent with the known chaperone function of Hsp90, although direct binding of Hsp90 to a client protein in the absence of Hsp40 and/or Hsc70 co-chaperones is unusual. The continued association of Hsp90 with its refolded client protein is also unusual. In fact, we could not identify any cellular factor that would displace Hsp90 from its CTA1 binding partner [29]. We accordingly proposed that Hsp90 remains bound to CTA1 after toxin translocation to the cytosol is complete, although technical limitations have to date prevented the detection of a cytosolic Hsp90/CTA1 complex.

The extended association of CTA1 with Hsp90 is likely beneficial for intoxication, as the loss of its stabilizing interaction with Hsp90 would return CTA1 to a disordered, inactive conformation. Consistent with this possibility, we have shown Hsp90 can impact host-toxin interactions that affect the enzymatic activity of CTA1. The free CTA1 polypeptide has little to no enzymatic activity at 37 °C because of its unstable nature [32]. However, the interaction between Hsp90/ATP and CTA1 at 37 °C produces both a gain-of-structure and gain-of-function for disordered CTA1. The activity of Hsp90-refolded CTA1 can be further enhanced by the addition of ADP-ribosylation factor 6 (ARF6) [29]. ARF proteins are allosteric activators of CTA1 and are essential for in vivo toxin activity, but ARFs do not induce a gain-of-structure in disordered CTA1 and can only stimulate the activity of a folded CTA1 polypeptide [33].

Lipid rafts, which are discrete cholesterol- and glycosphingolipid-enriched regions of the plasma membrane where Gsα is located, exhibit a "lipochaperone" property that also induces a gain-of-structure and gain-of-function in disordered CTA1 [34]. ARF6 can further enhance the basal enzymatic activity of raft-refolded CTA1 at 37 °C. Yet the greatest level of in vitro CTA1 activity at 37 °C is obtained with an initial exposure of disordered CTA1 to Hsp90/ATP, followed by a secondary addition of lipid rafts and ARF6. Surprisingly, the simultaneous addition of Hsp90/ATP, lipid rafts, and ARF6 to disordered CTA1 at 37 °C generates an enzymatically inactive toxin [29]. The order of host interactions with CTA1 thus seems crucial for toxin activity in the host cytosol. These collective observations are consistent with the cellular sequence of events in CT intoxication. As CTA1 emerges from the translocon pore, it contacts Hsp90 and apparently remains associated with the chaperone after extraction from the ER. The CTA1/Hsp90 complex then encounters Gsα in a lipid raft subdomain of the plasma membrane and utilizes an interaction with ARF proteins to further stimulate the ADP-ribosyltransferase activity of CTA1. The continued, post-translocation association of CTA1 with Hsp90 thus maintains CTA1 in a folded conformation and promotes its optimal level of enzymatic activity via subsequent interactions with lipid rafts and ARF proteins.

Hsp90 and Other AB Toxins

Hsp90 is required for the endosome-to-cytosol export of AB toxins with A chains that act as ADP-ribosyltransferases, but it is not involved with the endosomal translocation of AB toxins with A chains containing other enzymatic activities. It has therefore been proposed that Hsp90 specifically interacts with toxin A chains that are ADP-ribosyltransferases [35]. The interactions we have documented between Hsp90 and CTA1 are consistent with this model and have expanded the model to include toxin translocation from the ER as well as acidified endosomes. Furthermore, the mechanistic basis for Hsp90-mediated toxin extraction from the ER could explain the role of Hsp90 in toxin translocation across the endosomal membrane as well: in both cases, refolding of the toxin A chain by Hsp90 would prevent toxin back-sliding into the membrane pore and would thereby ensure unidirectional translocation to the cytosol.

The inhibition of Hsp90 by GA generates cellular resistance to several, but not all, AB toxins. We found GA treatment inhibits the cytotoxic effect of *Pseudomonas aeruginosa* exotoxin A, an AB-type protein toxin with an A chain that moves from the ER to the cytosol and functions as an ADP-ribosyltransferase (Fig. 5a). The protective effect of GA against exotoxin A was therefore consistent with the

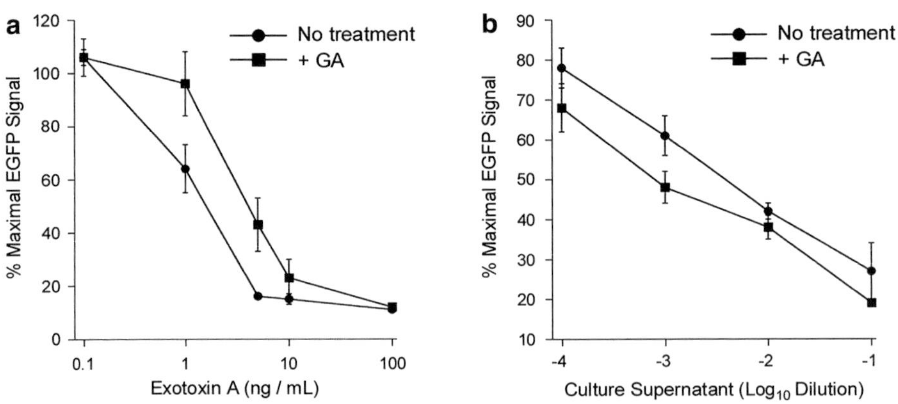

Fig. 5 Inhibition of Hsp90 specifically blocks the activity of an AB-type, ADP-ribosylating toxin. Vero cells expressing a destabilized variant of EGFP ($t_{½}=2$ h) were incubated 18 h in the absence or presence of 0.1 μM GA with varying concentrations of (**a**) exotoxin A or (**b**) Shiga toxin. As detailed in Massey et al. [36], continual synthesis of the destabilized EGFP is required for the Vero-d2EGFP cells to maintain a fluorescent signal. Exposure to exotoxin A or Shiga toxin inhibits protein synthesis and consequently generates a dose-dependent loss of the EGFP signal. Fluorescence was measured with a plate reader, and values from intoxicated cells were expressed as percentages of the values obtained from unintoxicated control cells. Shiga toxin 1 and Shiga toxin 2 were delivered to the Vero-d2EGFP cells in tenfold dilutions of cell-free culture supernatant from toxigenic *Escherichia coli* strain RM1697. The averages ± standard deviations of three independent experiments with six replicate samples per condition are shown. At the ED_{50} value, GA-treated cells were twofold more resistant to exotoxin A and eightfold more sensitive to Shiga toxin than toxin-challenged cells incubated without GA

Hps90-driven translocation of ADP-ribosyltransferases. Yet not every ER-translocating toxin follows an Hsp90-dependent exit route from the ER: GA treatment results in sensitization, rather than resistance, to Shiga toxin (Fig. 5b) and the plant toxin ricin [37]. In the case of ricin, toxin sensitization (which could only occur if ricin A chain was exported to the cytosol) was attributed to the inability of GA-treated cells to degrade the translocated pool of ricin A chain [37]. Neither Shiga toxin nor ricin contain an A chain ADP-ribosyltransferase. The initial suggestion that Hsp90 is a common factor in the cytosolic translocation of A chain ADP-ribosyltransferases has thus proven correct for all AB toxins tested to date, which strongly suggests Hsp90 recognizes a common feature of ADP-ribosylating toxins.

Summary

The CT intoxication process involves what has been termed an order-disorder-order transition [38]. CTA1 moves by vesicle carriers from the cell surface to the ER as a stable, ordered component of the CT holotoxin. In the ER, the PDI-mediated separation of CTA1 from CTA2/CTB$_5$ releases the structural constraints on CTA1 unfolding. The dissociated CTA1 subunit spontaneously shifts to a disordered conformation that is identified as a substrate for ERAD-directed export to the cytosol. CTA1 then interacts with factors in the host cytosol to regain an ordered, active conformation for the ADP-ribosylation of Gsα. As we have outlined here, Hsp90 plays a key role in the disorder-to-order transition that has functional consequences for both toxin translocation from the ER and toxin activity in the cytosol. Hs90 recognizes the N-terminus of disordered CTA1 as it emerges at the cytosolic face of the ER membrane and appears to couple CTA1 refolding with CTA1 extraction from the ER. The continued association of CTA1 with Hsp90 maintains the toxin in an ordered conformation for optimal, ARF-stimulated enzymatic activity. Our structure/function analysis has thus documented an essential role for the eukaryotic chaperone Hsp90 in the cell biology of cholera intoxication, with potential implications for therapeutic development and the cell biology of other AB-type, ADP-ribosylating toxins.

Acknowledgments Research in the Teter lab on CTA1-Hsp90 interactions has been supported by the National Institute of Allergy and Infectious Diseases of the National Institutes of Health under award numbers R01AI073783 and R01AI099493. The content is solely the responsibility of the authors and does not necessarily represent the official views of the National Institutes of Health.

References

1. Zhang RG, Scott DL, Westbrook ML, Nance S, Spangler BD, Shipley GG, Westbrook EM (1995) The three-dimensional crystal structure of cholera toxin. J Mol Biol 251:563–573
2. De Haan L, Hirst TR (2004) Cholera toxin: a paradigm for multi-functional engagement of cellular mechanisms (Review). Mol Membr Biol 21:77–92

3. Spangler BD (1992) Structure and function of cholera toxin and the related *Escherichia coli* heat-labile enterotoxin. Microbiol Rev 56:622–647
4. Kopic S, Geibel JP (2010) Toxin mediated diarrhea in the 21st century: the pathophysiology of intestinal ion transport in the course of ETEC, *V. cholerae* and rotavirus infection. Toxins 2:2132–2157
5. Harris JB, LaRocque RC, Qadri F, Ryan ET, Calderwood SB (2012) Cholera. Lancet 379:2466–2476
6. Wernick NLB, Chinnapen DJ-F, Cho JA, Lencer WI (2010) Cholera toxin: an intracellular journey into the cytosol by way of the endoplasmic reticulum. Toxins 2:310–325
7. Mekalanos JJ, Collier RJ, Romig WR (1979) Enzymic activity of cholera toxin. II. Relationships to proteolytic processing, disulfide bond reduction, and subunit composition. J Biol Chem 254:5855–5861
8. Khan HA, Mutus B (2014) Protein disulfide isomerase: a multifunctional protein with multiple physiological roles. Front Chem 2:1–9
9. Taylor M, Burress H, Banerjee T, Ray S, Curtis D, Tatulian SA, Teter K (2014) Substrate-induced unfolding of protein disulfide isomerase displaces the cholera toxin A1 subunit from its holotoxin. PLoS Pathog 10:e1003925
10. Taylor M, Banerjee T, Ray S, Tatulian SA, Teter K (2011) Protein disulfide isomerase displaces the cholera toxin A1 subunit from the holotoxin without unfolding the A1 subunit. J Biol Chem 286:22090–22100
11. Pande AH, Scaglione P, Taylor M, Nemec KN, Tuthill S, Moe D, Holmes RK, Tatulian SA, Teter K (2007) Conformational instability of the cholera toxin A1 polypeptide. J Mol Biol 374:1114–1128
12. Guerriero CJ, Brodsky JL (2012) The delicate balance between secreted protein folding and endoplasmic reticulum-associated degradation in human physiology. Physiol Rev 92: 537–576
13. Hazes B, Read RJ (1997) Accumulating evidence suggests that several AB-toxins subvert the endoplasmic reticulum-associated protein degradation pathway to enter target cells. Biochemistry 36:11051–11054
14. Teter K (2013) Cholera toxin interactions with host cell stress proteins. In: Henderson B (ed) Moonlighting cell stress proteins in microbial infections. Springer, New York, pp 323–338
15. Rodighiero C, Tsai B, Rapoport TA, Lencer WI (2002) Role of ubiquitination in retro-translocation of cholera toxin and escape of cytosolic degradation. EMBO Rep 3:1222–1227
16. Lemus L, Goder V (2014) Regulation of endoplasmic reticulum-associated protein degradation (ERAD) by ubiquitin. Cells 3:824–847
17. Kothe M, Ye Y, Wagner JS, De Luca HE, Kern E, Rapoport TA, Lencer WI (2005) Role of p97 AAA-ATPase in the retrotranslocation of the cholera toxin A1 chain, a non-ubiquitinated substrate. J Biol Chem 280:28127–28132
18. McConnell E, Lass A, Wojcik C (2007) Ufd1-Npl4 is a negative regulator of cholera toxin retrotranslocation. Biochem Biophys Res Commun 355:1087–1090
19. Zuehlke A, Johnson JL (2010) Hsp90 and co-chaperones twist the functions of diverse client proteins. Biopolymers 93:211–217
20. Solit DB, Chiosis G (2008) Development and application of Hsp90 inhibitors. Drug Discov Today 13:38–43
21. Haug G, Leemhuis J, Tiemann D, Meyer DK, Aktories K, Barth H (2003) The host cell chaperone Hsp90 is essential for translocation of the binary *Clostridium botulinum* C2 toxin into the cytosol. J Biol Chem 278:32266–32274
22. Ratts R, Zeng H, Berg EA, Blue C, McComb ME, Costello CE, vanderSpek JC, Murphy JR (2003) The cytosolic entry of diphtheria toxin catalytic domain requires a host cell cytosolic translocation factor complex. J Cell Biol 160:1139–1150
23. Wang X, Venable J, LaPointe P, Hutt DM, Koulov AV, Coppinger J, Gurkan C, Kellner W, Matteson J, Plutner H, Riordan JR, Kelly JW, Yates JR 3rd, Balch WE (2006) Hsp90 cochaperone Aha1 downregulation rescues misfolding of CFTR in cystic fibrosis. Cell 127:803–815

24. Youker RT, Walsh P, Beilharz T, Lithgow T, Brodsky JL (2004) Distinct roles for the Hsp40 and Hsp90 molecular chaperones during cystic fibrosis transmembrane conductance regulator degradation in yeast. Mol Biol Cell 15:4787–4797
25. Giodini A, Cresswell P (2008) Hsp90-mediated cytosolic refolding of exogenous proteins internalized by dendritic cells. EMBO J 27:201–211
26. Taylor M, Navarro-Garcia F, Huerta J, Burress H, Massey S, Ireton K, Teter K (2010) Hsp90 is required for transfer of the cholera toxin A1 subunit from the endoplasmic reticulum to the cytosol. J Biol Chem 285:31261–31267
27. Taylor M, Banerjee T, VanBennekom N, Teter K (2012) Detection of toxin translocation into the host cytosol by surface plasmon resonance. J Vis Exp (59): e3686. doi:10.3791/3686
28. Lencer WI, de Almeida JB, Moe S, Stow JL, Ausiello DA, Madara JL (1993) Entry of cholera toxin into polarized human intestinal epithelial cells. Identification of an early brefeldin A sensitive event required for A1-peptide generation. J Clin Invest 92:2941–2951
29. Burress H, Taylor M, Banerjee T, Tatulian SA, Teter K (2014) Co- and post-translocation roles for Hsp90 in cholera intoxication. J Biol Chem 289:33644–33654
30. Peskin CS, Odell GM, Oster GF (1993) Cellular motions and thermal fluctuations: the Brownian ratchet. Biophys J 65:316–324
31. Teter K, Jobling MG, Sentz D, Holmes RK (2006) The cholera toxin A13 subdomain is essential for interaction with ADP-ribosylation factor 6 and full toxic activity but is not required for translocation from the endoplasmic reticulum to the cytosol. Infect Immun 74:2259–2267
32. Murayama T, Tsai SC, Adamik R, Moss J, Vaughan M (1993) Effects of temperature on ADP-ribosylation factor stimulation of cholera toxin activity. Biochemistry 32:561–566
33. Banerjee T, Taylor M, Jobling MG, Burress H, Yang Z, Serrano A, Holmes RK, Tatulian SA, Teter K (2014) ADP-ribosylation factor 6 acts as an allosteric activator for the folded but not disordered cholera toxin A1 polypeptide. Mol Microbiol 94:898–912
34. Ray S, Taylor M, Banerjee T, Tatulian SA, Teter K (2012) Lipid rafts alter the stability and activity of the cholera toxin A1 subunit. J Biol Chem 287:30395–30405
35. Barth H (2011) Exploring the role of host cell chaperones/PPIases during cellular up-take of bacterial ADP-ribosylating toxins as basis for novel pharmacological strategies to protect mammalian cells against these virulence factors. Naunyn Schmiedebergs Arch Pharmacol 383:237–245
36. Massey S, Quinones B, Teter K (2011) A cell-based fluorescent assay to detect the activity of Shiga toxin and other toxins that inhibit protein synthesis. Methods Mol Biol 739:49–59
37. Spooner RA, Hart PJ, Cook JP, Pietroni P, Rogon C, Hohfeld J, Roberts LM, Lord JM (2008) Cytosolic chaperones influence the fate of a toxin dislocated from the endoplasmic reticulum. Proc Natl Acad Sci U S A 105:17408–17413
38. Ampapathi RS, Creath AL, Lou DI, Craft JW Jr, Blanke SR, Legge GB (2008) Order-disorder-order transitions mediate the activation of cholera toxin. J Mol Biol 377:748–760

Function, Structure and Stability of Human Gamma D Crystallins: A Review

Lina Rivillas-Acevedo, Arline Fernández-Silva, and Carlos Amero

Abstract The human γD crystallins is one of the most abundant protein of the lens nucleus and it is believed that their main function is to help maintaining the optical properties of the lens during the life span. Human γ-D crystallins is a 173 residues protein that fold into two homologousdomains (N-terminal and C-terminal), each containing two Greek key motifs. Although, this protein is extremely stable, over the years the protection mechanism loses efficiency and the protein accumulates damages, resulting in protein aggregation which is associated with cataracts formation. At present, cataracts is the main cause of blindness in the world and surgical removal of the lens remains the only treatment. This review summarizes the current knowledge on cataracts risk factors, in vivo studies, unfolding and inhibition associated with human γD crystallins.

Keywords Human γD crystallin • Cataracts • Blindness • Protein unfolding • Aggregates • Inhibition • Greek key motifs • Protein damage • Blindness • *In vivo* studies

Introduction

The main function of the eye lens is to produce the refractive power to focus the light onto the retina [1]. Lens is an avascular tissue composed of a single layer of epithelial cells, which has to maintain the transparency to allow the passage of light [2]. During the lens formation, the epithelial cells elongate to become fiber cells.

L. Rivillas-Acevedo • A. Fernández-Silva • C. Amero (✉)
Laboratorio de Bioquímica y Resonancia Magnética Nuclear,
Centro de Investigaciones Químicas, Instituto de Investigación en Ciencias Básicas y Aplicadas, Universidad Autónoma del Estado de Morelos, Avenida Universidad 1001, Colonia Chamilpa, Cuernavaca, Morelos 62209, Mexico
e-mail: carlosamero@uaem.mx

L. Olivares-Quiroz et al. (eds.), *Physical Biology of Proteins and Peptides*,
DOI 10.1007/978-3-319-21687-4_5

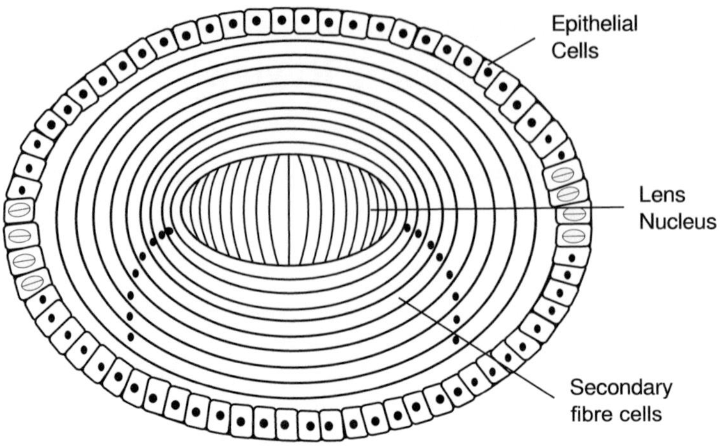

Fig. 1 Lens morphology. Lens nucleus is formed by the primary fiber cells, which differentiated from epithelial cells during embryogenesis. After the elongation, the organelles (*black dots*) are eliminated from the fiber cells to avoid light scattering

In this process of differentiation, the cells produce large quantities of proteins. After this elongation, in order to avoid any potential light scattering structure, the organelles are removed from the fiber cells by a programmed proteolytic process (Fig. 1) [3, 4]. Even though this process continues throughout life, it does at a minimal pace, and the older fibers are pushed to the center of the lens by the newer fiber cells (Fig. 1) [2]. Therefore, fully differentiated lens cells consist of a high concentration of proteins without nuclei or any other organelles and do not produce or degrade proteins [1]. Human lens has the highest protein content within the body [5] and, as a direct consequence of the missing repair mechanisms, the proteins in the lens have to be stable and soluble for the organism lifetime.

Most of the refractive power of the lens is conferred by high concentrations of three types of proteins: α, β and γ-crystallins, which count for more than 90 % of the proteins in the lens [6, 7]. In normal lenses, these proteins are found in concentrations higher than 300 mg/mL in order to have a high refractive index [6], and while they still scatter light, they do it in a specific manner. The correlation in the position of pairs of proteins reduces the scattering; this is due to the fact that the fluctuation in the number of protein molecules over a short space is small, resulting in transparency [8]. This is known as short-range spatial order. Any microscopic fluctuation will induce light scattering, these fluctuations could be produced by molecules of large molecular weight formed, for example, by protein aggregation or polymerization [8]. Such protein aggregation leads to the development of opacities known as cataracts.

Cataracts Risk Factors

The cataracts are an opacification of the lens of the eyes due to the accumulation of aggregates of crystallins. At present, cataracts formation is the main cause of blindness in the world [9]. The formation of cataracts is highly related to aging, and it is generally accepted that proteins modifications have to occur to promote the aggregation and then the cataracts. Some of the modifications that have been observed with aging, include oxidation of the tryptophans, cysteines and methionines [10], deamidation of arginines and glutamines [11, 12], glycosylation of lysine residues [13], and peptide fragmentation [14]. Most of these modifications are believed to be induced by environmental factors such as ultraviolet light exposure [15, 16], and oxidative stress [10, 17, 18]; the consequence for all of them is the formation of insoluble aggregates resulting in lens opacity and then loss of vision for cataracts (Table 1).

In the human eye, the cornea absorbs most of the UV radiation (240–400 nm) [27], and there is also low-molecular weight UV-filter molecules in the human lens, the most abundant are kynurenine, 3-hydroxykynurenine glucoside (3OHKG), and 4-(2-amino-3-hydroxyphenyl)-4-oxybutanoic acid O-glucoside, which absorb in the UV-B (280–330 nm) and UV-C ranges (100–280 nm) [28]. However, throughout the years, significant radiation reaches the lens, producing proteins UV radiation damage, which has been correlated manly to structural changes in different residues that destabilize the proteins and accelerate their unfolding [16].

Table 1 Some of the rick factor and modification associated with cataracts

Risk factor	Modification	Damage	Reference
Aging	Deamidation	Introduce carboxilate anion, which induce ionic interactions that may lead to conformational changes in the protein	Stacy et al. [19]
	Methylation	Avoids disulfide bond formation preventing aggregation	Lapko et al. [20]
	Disulphide bonds	Formation of intra-chain disulfide bonds causing protein conformational changes	Stacy et al. [19]
		Multimer formation by inter-chain disulfide bonds	
UV radiation	Photodamage reaction of tryptophan residues	Destabilization of the native structure of the protein	Borkman [21]
Mutations	R14C	Altered protein surface properties	Stephan et al. [22]
	P23T	Reduced solubility	Mackay et al. [23]
	R36S	Crystallisation	Kmoch et al. [24]
	R58H	Impaired protein folding	He et al. [25]
	W158X	Fourth Greek key motif	Gopinath et al. [26]

On the other hand, the oxidative stress becomes more important with aging, due to the decline of the levels of glutathione (GSH) and antioxidant enzymes as the superoxide dismutase and catalase activity [29, 30]. The decrease of GSH levels lead to crosslinking, aggregation, insolubility, and fragmentation of crystallin, which then result in the formation of cataracts [31]. Moreover, aging may also reduce the functionality of the chaperon α-crystallins to rescue the unfold β and γ-crystallins [32].

As well as aging, diabetes, malnutrition, hypertension and ionic imbalance have been correlated with cataracts formation. For instance, in diabetes, the hyperglycemia produced the diffusion of extracellular glucose to lens which produced then post-translational modification in the proteins of the lenses [33]. Also, the glucose is metabolized by the sorbitol-aldose reductase pathway and, as a result, the lens accumulates sorbitol. Then the sorbitol produces osmotic swelling, leakage of glutathione, myo-inositol changes in the membrane permeability, and generation of free radicals [34, 35]. Another factor, associated with cataracts is the ionic imbalance in the lens. An increase in the concentrations of calcium produces the activation of calpain serine-proteases that will degrade crystallins [36, 37], while a magnesium concentration decrease has been associated with an increase in the oxidative stress [38].

Finally, an intrinsically different group of cataracts are the hereditary cataracts which, unlike the other types, occur at early ages. The congenital cataracts are caused by a single-gene disorder that induces the substitution of one amino-acid, and the estimated frequency is between 1 and 6 per 10,000 of new born babies [39]. Some of these mutations are listed in Table 1; the mutation can cause protein insolubility, crystallization, changes in charge, or destabilization.

Even tough there are several factors related to cataract formation, the only effective treatment is still surgical removal of the lens. This procedure has become routine in many clinics, but nevertheless cataracts remain as the main cause of blindness in the world. This is due to the fact that this surgical procedure is expensive and is often inaccessible in low and middle-income countries. Also, it has been argued that the artificial lens doesn't have all the optical properties of the "normal" lens [40, 41].

In Vivo Studies

The use of *in vivo* studies allows to conduct essential investigations in the characteristic physiological environment of the disease, providing a deeper understanding of the mechanisms involved in the appearance and progression of cataracts. Therefore, several different animals models have been developed, capable of reproducing the characteristics of the disease seen in humans, mainly to study the different factors that give rise to cataracts and some potential treatment molecules. These studies would contribute decisively to the possible development of future preventive treatments.

Some of the earliest studies focused on radiation damage induced in lenses. Rabbit lenses were subjected to X-rays, and by electronic microscopy, it was possible to observe very high molecular aggregates in irradiated cortex while in normal cortex only relatively small aggregates were visible [119]. Later, some studies of UV radiation effects in rodents lenses were done, demonstrating by histological analysis that the first target tissue of UV radiation was the epithelium [42]. Also it was found that the level of glutathione (GSH) suffers modification [43]. With these types of experiments, it has been generally accepted that the long term effect of UV-radiation is cataract formation probably by protein damage [44].

Studies focusing in oxidative stress have made extensive use of specific mouse models with known alterations for oxidative damage, and were then able to reproduce some of the features of human aged related cataracts [45, 46]. For instance, the OXXYS strain exhibits premature aging and significantly shortened life span associated with oxidative damage, and therefore the status of the lens can be characterized due to the inherent over-generation of ROS (Radical Oxygen Species) [47]. In a different methodology, oxidative stress can be induced in rats trough a galactose diet, and then induce the depletion of levels of GSH [48].

Cataract has also been studied in animal model using other inducers. Selenite has been commonly used, causing alteration of the epithelial metabolism, calcium accumulation, calpain-induced proteolysis, crystallins precipitation and cytoskeletal [49, 50]. Galactose-induced cataracts produce the accumulation of osmolytes causing ocular lens swelling of fiber cells and then disruption of the native state of lens proteins [51]. Cataracts have also been studied by using genetics mouse models [45, 52, 53].

In a different type of studies, the specific aim of using animal models has been to test routes of administration of compounds, like intraperitoneal [29, 30, 54], subcutaneous [55] or as a supplementary diet [56, 57].

Crystallins Family

Crystallins are the main constituent of the lens and are separated into three types in vertebrates: α, β and γ-crystallins [6, 7]. α-Crystallin is the predominant and biggest type of crystallins, with a molecular weight around 800 kDa [6]. It is formed by the non-covalent association of two subunits (A and B), in a molar ratio around 3:1 to 2:1 [58, 59]. Subunits A and B are homologue and form part of the small heat shock proteins family (sHSPs). Their secondary structure is primarily β-sheets [60], and the subunits seem to be in equilibrium between dissociating and re-associating constantly, which makes the protein a dynamic oligomer [61]. The main function of the native α-crystallin complex in the lens is its chaperone function [60, 62–65], but also, some experimental evidence suggests that α-crystallin is involved in remodeling and protection of the cytoskeleton, inhibition of apoptosis and the resistance to stress [66].

The β and γ-crystallins are members of the same family; they have 30 % sequence identity [67, 68], consisting mainly of β-sheets that form two domains each formed for two Greek Key motif. The β-crystallins are multimeric and contain N-terminal extensions, while the γ-crystallins are monomeric in solution [6]. The β-crystallins family is constituted by four acidic proteins (βA1, βA2, βA3, and βA4) and three basic (βB1, βB2, βB3) [69, 70]. The βB1, βB3 and βA4 crystallins are homo-oligomers [71, 72] while the βB2-crystallin forms dimers by domain swapping, and hetero-oligomers with βA1, βA2 and βA3-crystallin. The domain swapped dimer is thermodynamically unstable [67, 73], but has been proposed to act as a chaperon to stabilize or to assemble other β-crystallins [74]. The γ-crystallins are encoded by seven different genes, γA, γB, γC, γD, γE, γF, and γS, all encoding proteins of about 20 kDa. Nevertheless, only the γD, γC and γS crystallins are expressed at an appreciable concentration in human lenses [66].

Human γD-Crystallin Structure and Function

The human γD-crystallin (HγD) is the second most abundant protein of the lens nucleus (Ji 2013), and it is believed that their main function is to help maintaining the optical properties of the lens during the life span. Consequently, the γD-crystallins have evolved to be stable, soluble at elevated concentrations and damage tolerant.

HγD is a monomeric protein with 173 amino acids. It is composed of two domains: N and C-terminal domain connected by a short linker (Fig. 2). Its structure has been resolved by X-ray diffraction [75]. It folds into two Greek key motifs with eight β-strands characterized by an unusual folded β-hairpin (Fig. 2).

Some of the features that seem to confer stability to the protein include six pair of aromatic residues distributed all over the protein (F11/Y6, Y16/Y28, Y50/Y45, Y92/Y97, Y138/y133, F117/F115) [76], a group of hydrophobic and polar residues in the inter-face of the two domains [77] and two "tyrosine corners", a classic characteristic of the Greek key motifs [78]. This motif is formed in the N-terminal domain with tyrosine 62 and in the C-terminal domain with tyrosine 150, and it connects two β-strands by a hydrogen bond between the hydroxyl group of the tyrosine and a carboxyl group [79]. Interestingly, HγD have six cysteines, more than any of the α and β-crystallins, but does not have any disulphide bond [80].

While the UV-damage protection is conferred by four buried tryptophan: W42 and W68 in the N-terminal domain, and W130 and W156 in the C-terminal domain, these tryptophan are highly conserved among vertebrates [6] and are important in the stability of the protein by dispersing the UV radiation [16, 81]. The conformation of the tryptophans seems to enable the lens to be a very effective UV filter, and their efficient quenching of the radiation provides an excellent mechanism to prevent photochemical degradation [82].

Although, this protein is extremely stable, over the years the protection mechanism loses efficiency and the protein accumulates damages. Some of the post-translational modifications of γD-crystallins include deamidation of Q12, N49 and

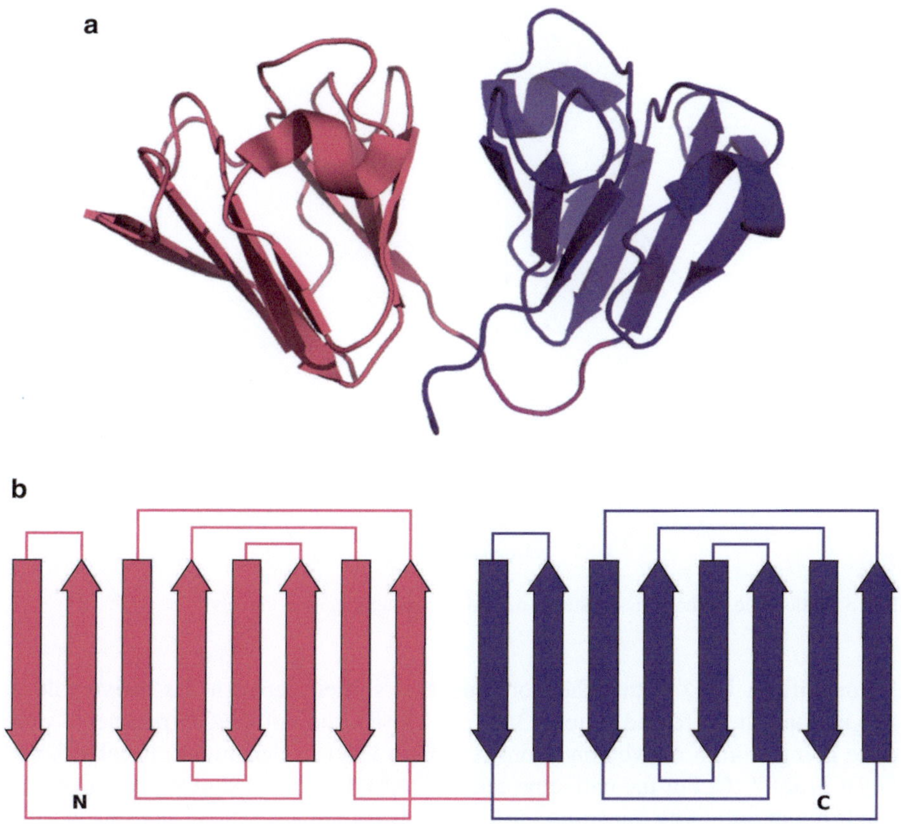

Fig. 2 HγD-crystallin structure. (**a**) Three-dimensional structure of HγD-crystallin (2KFB), drawn in Pymol. (**b**) Topology of the N-terminal domain (*pink*), and C-terminal domain (*purple*)

N160 [83, 84], Y45 and W156 oxidation [83, 85], C110 methylation [20], G1 carbamylation [20, 84], and C18-C32 and C108-C110 disulfide binding [11]. All of these major modifications become more important with aging, and are considered to be the first step in cataract development [12].

Unfolding: Aggregation

As in many other protein folding diseases, it is believed that a partially unfolded intermediate produces the formation of the aggregates that lead to cataracts. Consequently, the aggregation pathway for the HγD-crystallin has been extensively studied. By using equilibrium temperature and chemical unfolding, it was found that the HγD-crystallin is stable to 8 M urea and has a Tm close to 80 °C. In order to unfold the protein it was necessary to use GuHCl at 37 °C [86]. A very useful

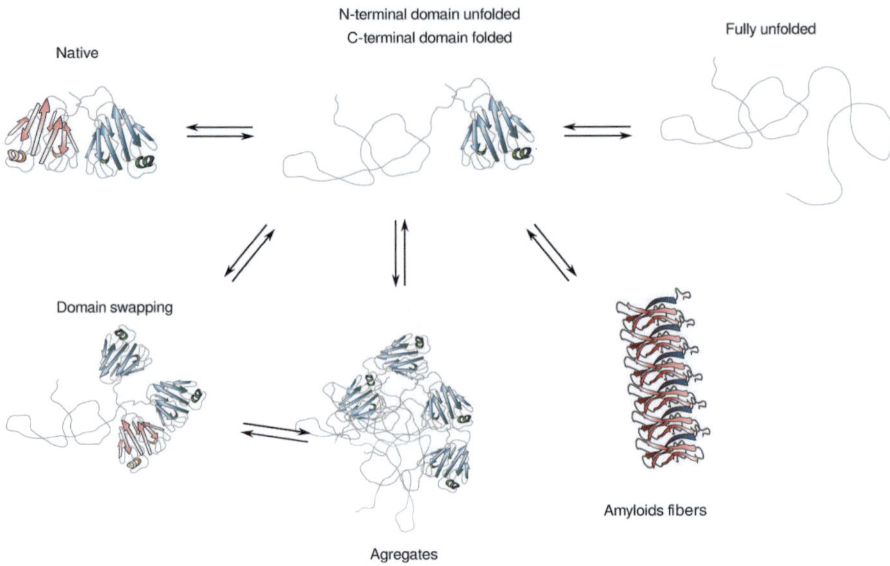

Fig. 3 Schematic of the HγD-crystallin unfolding and aggregation models

feature of the HγD is that the florescence is shifted between the native fold at 325 nm and the unfolded state at 348 nm. Mostly, the unfolded experiments were done going from a native conditions to a high level of denaturant, usually 5.5 M GuHCL at 37 °C. For the native protein the unfolding curve showed, apparently, a single major unfolding transition [86]; later it was found by kinetic analysis in triple mutants the existence of a partially unfolded intermediate [87], and it was confirmed that the native unfolding curve was better adjusted by a three state model [88]. The first unfolding event consists in the unfolding of the N-terminal domain, intermediately followed by the unfolding of the C-terminal domain. Even though the state with one domain structured and the other one unfolded is a short live state in the native, it has been proposed as a good candidate for the aggregation intermediate in the cataract formation (Fig. 3).

There has been several stability/unfolding studies of mutants proteins, mainly divided in two types: mutations to change the stability elements of the protein and mutations that mimic some protein damage observed in *in vivo* experiments. In the first group, there are reports for the mutations of the conserved aromatic pairs in the protein F11/Y6, Y16/Y28, Y50/Y45, Y92/Y97, Y138/Y133, F117/F115 [76], mutations of the four internal tryptophans W42,W68/W130/W156, mutations of residues in the protein interface (Q54, Q143, and R79, M147) [88] among others. Meanwhile, in the second group, most of the mutations evaluated have been a single mutation that has been reported in cataracts studies. For instance, the genetic studies of families exhibiting juvenile-onset cataracts R114, R38, and R36 [24, 25], possible glutamine deamidation sites Q54, Q143 [89], deletion mutant W156X [90], R140, A36 [91], among others.

Most of the mutations can be classified according to where the mutation occurred: the N-domain, the interface or the C-domain. In general, N-terminal domain mutations shifted the first transition to lower denaturan concentration without affecting the C-terminal domain. The lower stability of the first unfolding event, in many cases, populated more this intermediate state and clearly presented a three unfolding curve [76], whereas mutation in the C-domain did not usually affect the first unfolding event or the shape of the curve. The mutations seems to destabilize the two domains in a more cooperative way [76].

Finally, the mutations in the interface produce a destabilization of the N-terminal and reveal a clear plateau in the unfolding curves [88, 89]. And, at the same time, they affect the N-terminal domain folding but not the C-terminal domain, suggesting that the interface could function as a scaffold for the folding of the N-terminal domain [88]. Many of these mutations studied would, consequently, effectively increase the life span of the partially unfolded conformer. Interestingly, many of the observed cataract point mutations *in vivo* cluster in the N-terminal domain.

It has been also studied the unfolding of the protein due to external factors. For instance, by using UV radiation, depending on the exposition time, protein aggregation has been observed [1, 92], as well as several damages from affecting specific residues to complete cleavage of the backbone [93]. It is worth to notice that out of the four internal tryptophan, it was reported that W42 is the more easily affected by UV radiations [94]. As a different external factor, the protein has been exposed to acidic pHs, and the formation of amyloid fibril was observed in these conditions (Fig. 3) [95].

Ultimately, the unfolding processes have been studied by molecular simulation *in silico*. In this type of studies, it has also been observed the preferential unfolding of the N-terminal domain, due to urea, temperature or mutations [96, 97]. Also it was proposed that the N-terminal domain stability is regulated by the inter-domain interaction. In simulations were the four internal tryptophan were changed to kynurenine simulating UV damage, it was found that both domain unfold faster and not sequentially. This suggests that, if the UV radiation is such that it makes this damage to any of the four tryptophans, the domain is going to unfold [16]. Also by simulation, the formation of a swap dimer was observed (Fig. 3) [96].

Inhibition

As mentioned before, cataracts is the main cause of blindness in the world [9], and while surgical removal of lens is effective, it is also expensive and is not available for all persons in developing countries. Hence, a search of a preventive treatment remains a key aspect in the research related to this disease.

Many different compounds have been studied in order to measure their anti-cataractogenic ability (Fig. 4). Several small molecules have been tested to prevent cataract formation due to external factors. For instance, caffeine was reported as an effective compound to prevent damage of lens by UV radiation [48]. The synthesis of bifunctional antioxidant with radical scavenging and chelation ability was found to delay the opacity of lenses exposed to gamma radiation and delay reactive oxygen species [56].

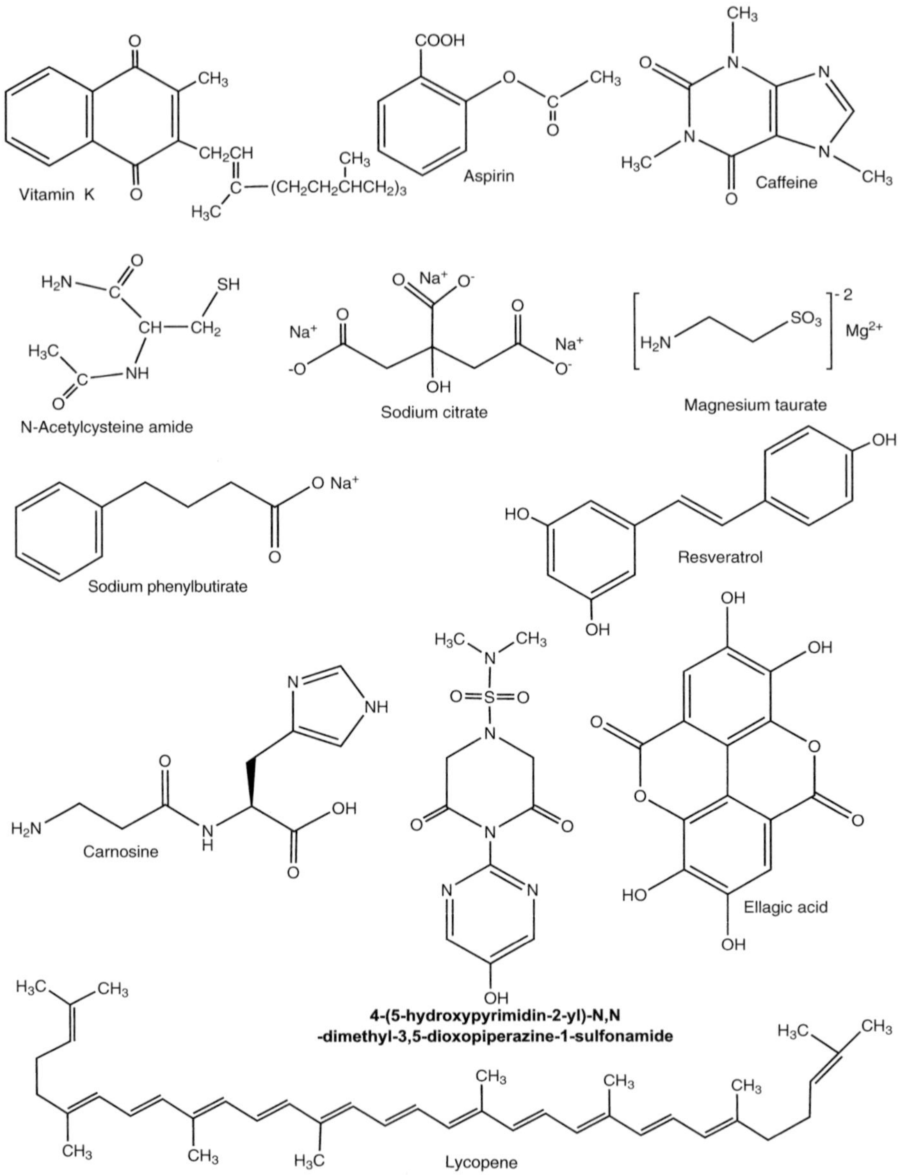

Fig. 4 Inhibitors. Chemical structure of some of the molecules that have been tested as cataract inhibitors

Nevertheless, most of the studies have focused in the inhibition of the factor responsible of producing cataracts. For instance, it has been reported that byakangelicin, the principal component of roots of *Angelica dahurics, suppress* sorbitol accumulation and galactose-induced cataract [98], citric acid inhibits the progression of cataracts in diabetics rat [99], and caffeine inhibits the oxidative stress in the lens [48].

In diabetic mouse models, there has been several reports of molecules studied to prevent cataracts. It was first tested the inhibitory effect of aspirin [100], then better results were obtained by combining aspirin and carnosine eye drops [101]. Vitamin K was showed to be effective to keep Ca^{2+} ATPase activity and lens calcium homeostasis resulting in an inhibition of diabetic cataractogenesis [57]. Similarly, magnesium taurate treatment to galactose diabetic rats showed its ability for correcting the lenticular Ca^{2+}/Mg^{2+} ratio [55]. Also, the aldose reductase has been a target for anti-cataract drug [31, 33].

There have been studies on inhibition of cataract formation by using several natural products, which are known to contain anti-oxidants (Table 2). The crude extract of plants *Ocimum sanctum* [110], *Ginkgo biloba* [112], and *Adhatoda vasica* [102], have been tested to inhibit the polyol accumulations, oxidative stress and aldose reductase activity in animal models. Flavonoid fractions of *Vitex negundo* present an increase effect in the enzymatic activity of superoxide dismutase, glutathione peroxidase and catalase [29]. Other molecules tested include flavonoids isolated from honey [113] and egalic acid [30].

Table 2 Plants tested to treat cataracts

Name of plants	Part of plant	Fraction	Mechanism of action	Reference
Morinda cetrifolia (Rubiaceae)	Leaves	Aqueous extract	Inhibition of aldose reductase	Gacche and Dhole [102]
Foeniculum vulgare (Apiaceae)	Fruit	Aqueous extract	Inhibition of aldose reductase	Dongare et al. [103]
Momordica charantia (Cucurbitaceae)	Fruit	Aqueous extract Ethanol extract	Antihiperglycemic effects in experimental diabetics	Rathi et al. [104]
Cheilanthes glauca (Adiantaceae)	Stems	Aqueous	Inhibits generated oxidative stress	Pastene et al. [105]
Brassica oleracea (Brassicaceae)	Flower	Flavonoids fraction	Inhibits selenite induced-cataracts	Vibin et al. [106]
Emblica officinalis (Phyllanthaceae)	Fruit	Aqueous extracts	Inhibits aldose reductase activity	Suryanarayana and Petrash [107]
Angelica dahurica (Apiaceae)	Roots	Ether extracts	Inhibits aldose reductase activity	Shin et al. [108]
Vitex negundo (*Verbenaceae*)	Leaves	Flavonoids	Inhibits oxidative stress	Rooban et al. [29]
Emilia sonchifolia (Asteraceae)	Leaves	Flavonoids fraction	Inhibits lens opacification and oxidative stress	Lija et al. [109]
Ocimum sanctum (*Lamiaceae*)	Plant materials	Aqueous extract	Inhibits aldose reductase	Halder et al. [110]
Cochlospermum religiosum (Cochlospermaceae)	Leaves	Isorhamnetin-3-glucoside fraction	Inhibits selenite induced-cataracts	Devi et al. [111]

Meanwhile, in studies *in vitro*, the search of cataract inhibition focuses in molecules that present an effect in the native γ-crystallin or the partially folded intermediaries. Making use of spectroscopic techniques, the ability of sodium citrate to interact with human wild type and mutants γD-crystallin was tested, showing that the molecule interacts with unfolding intermediates, and promotes stabilizing interactions [114]. While sodium 4-phenyl butyrate increments the solubility of a truncated γD-crystallin mutant [115] and pantethine stops the aggregation in solution and reduces the phase separation temperature [116]. For aggregation induced by UV light, resveratrol inhibits the aggregation [117] and carnosine inhibits the induced degradation [118].

Perspectives

Crystallins are not regenerated in the lens during the organism lifetime, therefore, they have to remain unchanged during aging in order to not affect the lens transparency. When different factors affect the protein, cataracts are formed. The understanding of the conformational changes that γ-crystallins suffer during the cataracts development is necessary in order to get structural information that would allow the development of treatment strategies different than the surgical removal. In spite of the multiple compounds that have been studied as potential drugs for cataracts, there is not a potential preventive molecule for the inhibition of aggregation of crystallins. The study of inhibitors previously related to or associated with other misfolding protein aggregation diseases may be a (valuable) option for testing new compounds.

References

1. Schafheimer N, Wang Z, Schey K, King J (2014) Tyrosine/cysteine cluster sensitizing human γD-crystallin to ultraviolet radiation-induced photoaggregation in vitro. Biochemistry 53:979–990. doi:10.1021/bi401397g
2. Sharma KK, Santhoshkumar P (2009) Lens aging: effects of crystallins. Biochim Biophys Acta 1790:1095–1108. doi:10.1016/j.bbagen.2009.05.008
3. Bassnett S (2002) Lens organelle degradation. Exp Eye Res 74:1–6. doi:10.1006/exer.2001.1111
4. Bassnett S (2009) On the mechanism of organelle degradation in the vertebrate lens. Exp Eye Res 88:133–139. doi:10.1016/j.exer.2008.08.017
5. Kyselova Z (2011) Mass spectrometry-based proteomics approaches applied in cataract research. Mass Spectrom Rev 30:1173–1184. doi:10.1002/mas.20317
6. Bloemendal H, de Jong W, Jaenicke R et al (2004) Ageing and vision: structure, stability and function of lens crystallins. Prog Biophys Mol Biol 86:407–485. doi:10.1016/j.pbiomolbio.2003.11.012
7. Slingsby C, Wistow GJ, Clark AR (2013) Evolution of crystallins for a role in the vertebrate eye lens. Protein Sci 22:367–380. doi:10.1002/pro.2229

8. Benedek GB (1971) Theory of transparency of the eye. Appl Opt 10:459–473. doi:10.1364/AO.10.000459
9. Fraile García E (2014) Causas de Ceguera en el Mundo: distribución geográfica y relación con el medio socio-económico
10. Hains PG, Truscott RJW (2007) Post-translational modifications in the nuclear region of young, aged, and cataract human lenses. J Proteome Res 6:3935–3943. doi:10.1021/pr070138h
11. Hanson SR, Hasan A, Smith DL, Smith JB (2000) The major in vivo modifications of the human water-insoluble lens crystallins are disulfide bonds, deamidation, methionine oxidation and backbone cleavage. Exp Eye Res 71:195–207. doi:http://dx.doi.org/10.1006/exer.2000.0868
12. Hanson SR, Smith DL, Smith JB (1998) Deamidation and disulfide bonding in human lens gamma-crystallins. Exp Eye Res 67:301–312. doi:10.1006/exer.1998.0530
13. Chiou SH, Chylack LT, Tung WH, Bunn HF (1981) Nonenzymatic glycosylation of bovine lens crystallins. Effect of aging. J Biol Chem 256:5176–5180
14. Ajaz MS, Ma Z, Smith DL, Smith JB (1997) Size of human lens β-crystallin aggregates are distinguished by N-terminal truncation of βB1. J Biol Chem 272:11250–11255. doi:10.1074/jbc.272.17.11250
15. Wang SS-S, Wen W-S (2010) Examining the influence of ultraviolet C irradiation on recombinant human γD-crystallin. Mol Vis 16:2777–2790
16. Xia Z, Yang Z, Huynh T et al (2013) UV-radiation induced disruption of dry-cavities in human γD-crystallin results in decreased stability and faster unfolding. Sci Rep 3:1560. doi:10.1038/srep01560
17. Moreau KL, King JA (2012) Cataract-causing defect of a mutant γ-crystallin proceeds through an aggregation pathway which bypasses recognition by the α-crystallin chaperone. PLoS One 7(5):e37256. doi:10.1371/journal.pone.0037256
18. Truscott RJW (2005) Age-related nuclear cataract-oxidation is the key. Exp Eye Res 80:709–725. doi:10.1016/j.exer.2004.12.007
19. Stacy RAH, David LS, Smith JB (1998) Deamidation and disulfide bonding in human lens γ-crystallins. Exp Eye Res 67:301–312
20. Lapko VN, Smith DL, Smith JB (2003) Methylation and carbamylation of human gamma-crystallins. Protein Sci 12:1762–1774. doi:10.1110/ps.0305403
21. Borkman F (1990) Mechanism of photochemically protein produced solutions in lens. Exp Eye Res 51:663–669
22. Stephan DA, Gillanders E, Vanderveen D et al (1999) Progressive juvenile-onset punctate cataracts caused by mutation of the γD-crystallin gene. Proc Natl Acad Sci U S A 96:1008–1012
23. Mackay DS, Andley UP, Shiels A (2004) A missense mutation in the γD crystallin gene (CRYGD) associated with autosomal dominant "coral-like" cataract linked to chromosome 2q. Mol Vis 10:155–162
24. Kmoch S, Brynda J, Asfaw B et al (2000) Link between a novel human gammaD-crystallin allele and a unique cataract phenotype explained by protein crystallography. Hum Mol Genet 9:1779–1786. doi:10.1093/hmg/9.12.1779
25. Héon E, Priston M, Schorderet DF et al (1999) The γ-crystallins and human cataracts: a puzzle made clearer. Am J Hum Genet 65:1261–1267. doi:10.1086/302619
26. Gopinath PM, Löster J, Graw J (2002) Novel mutations in the γ-crystallin genes cause autosomal dominant congenital cataracts. J Med Genet 39:352–358
27. Kolozsvári L, Nógrádi A, Hopp B, Bor Z (2002) UV absorbance of the human cornea in the 240- to 400-nm range. Invest Ophthalmol Vis Sci 43:2165–2168
28. Truscott RJ, Wood AM, Carver JA et al (1994) A new UV-filter compound in human lenses. FEBS Lett 348:173–176. doi:10.1016/0014-5793(94)00601-6
29. Rooban BN, Lija Y, Biju PG et al (2009) Vitex negundo attenuates calpain activation and cataractogenesis in selenite models. Exp Eye Res 88:575–582. doi:10.1016/j.exer.2008.11.020

30. Sakthivel M, Elanchezhian R, Ramesh E et al (2008) Prevention of selenite-induced cataractogenesis in Wistar rats by the polyphenol, ellagic acid. Exp Eye Res 86:251–259. doi:10.1016/j.exer.2007.10.016
31. Patel PM, Jivani N, Malaviya S et al (2012) Cataract: a major secondary diabetic complication. Int Curr Pharm J 1:180–185. doi:10.3329/icpj.v1i7.10813
32. Ponce A, Sorensen C, Takemoto L (2006) Role of short-range protein interactions in lens opacifications. Mol Vis 12:879–884. doi:v12/a99 [pii]
33. Esteves JF, Dal Pizzol MM, Sccoco CA et al (2008) Cataract and type 1 diabetes mellitus. Diabetes Res Clin Pract 82:324–328. doi:10.1016/j.diabres.2008.08.023
34. Jung HA, Islam MDN, Kwon YS et al (2011) Extraction and identification of three major aldose reductase inhibitors from Artemisia montana. Food Chem Toxicol 49:376–384. doi:10.1016/j.fct.2010.11.012
35. Stefek M (2011) Natural flavonoids as potential multifunctional agents in prevention of diabetic cataract. Interdiscip Toxicol 4:69–77. doi:10.2478/v10102-011-0013-y
36. Biswas S, Harris F, Dennison S et al (2004) Calpains: targets of cataract prevention? Trends Mol Med 10:78–84. doi:10.1016/j.molmed.2003.12.007
37. Dilsiz N, Olcucu ALI, Atas M (1999) Determination of calcium, sodium, potassium and magnesium concentrations in human senile cataractous lenses. Cell Biochem Funct 18:259–262
38. Nagai N, Fukuhata T, Ito Y (2007) Effect of magnesium deficiency on intracellular ATP levels in human lens epithelial cells. Biol Pharm Bull 30:6–10. doi:10.1248/bpb.30.6
39. Ji F, Jung J, Gronenborn AM (2012) Structural and biochemical characterization of the childhood cataract-associated R76S mutant of human γD-crystallin. Biochemistry 51:2588–2596. doi:10.1021/bi300199d
40. Hirsch RP, Schwartz B (1983) Increased mortality among elderly patients undergoing cataract extraction. Arch Ophthalmol 101:1034–1037. doi:10.1001/archopht.1983.01040020036004
41. Toh T, Morton J, Coxon J, Elder MJ (2007) Medical treatment of cataract. Clin Experiment Ophthalmol 35:664–671. doi:10.1111/j.1442-9071.2007.01559.x
42. Wu K, Shui YB, Kojima M et al (1997) Location and severity of UVB irradiation damage in the rat lens. Jpn J Ophthalmol 41:381–387
43. Wegener AR (1994) In vivo studies on the effect of UV-radiation on the eye lens in animals. Doc Ophthalmol 88:221–232. doi:10.1007/BF01203676
44. Zhang J, Yan H, Löfgren S et al (2012) Ultraviolet radiation-induced cataract in mice: the effect of age and the potential biochemical mechanism. Invest Ophthalmol Vis Sci 53:7276–7285. doi:10.1167/iovs.12-10482
45. Graw J (2009) Mouse models of cataract. J Genet 88:469–486. doi:10.1007/s12041-009-0066-2
46. Wolf N, Penn P, Pendergrass W et al (2005) Age-related cataract progression in five mouse models for anti-oxidant protection or hormonal influence. Exp Eye Res 81:276–285. doi:10.1016/j.exer.2005.01.024
47. Marsili S, Salganik RI, Albright CD et al (2004) Cataract formation in a strain of rats selected for high oxidative stress. Exp Eye Res 79:595–612. doi:10.1016/j.exer.2004.06.008
48. Varma SD, Hegde KR, Kovtun S (2010) Inhibition of selenite-induced cataract by caffeine. Acta Ophthalmol 88:245–249. doi:10.1111/j.1755-3768.2010.02014.x
49. Ostadalova I, Babický A, Odenberger J (1978) Cataract induced by administration of a single dose of sodium selenite to suckling rats. Experientia 15:222–223
50. Shearer TR, Ma H, Fukiage C, Azuma M (1997) Selenite nuclear cataract: review of the model. Mol Vis 3:8
51. Srivastava SK, Ramana KV, Bhatnagar A (2005) Role of aldose reductase and oxidative damage in diabetes and the consequent potential for therapeutic options. Endocr Rev 26:380–392. doi:10.1210/er.2004-0028
52. Liu XY, Dangel AW, Kelley RI et al (1999) The gene mutated in bare patches and striated mice encodes a novel 3β-hydroxysteroid dehydrogenase. Nat Genet 22:182–187

53. Sheets NL, Chauhan BK, Wawrousek E et al (2002) Cataract and lens specific upregulation of ARK receptor tyrosine kinase in emory mouse cataract. Invest Ophthalmol Vis Sci 43:1870–1875
54. Gao X, Wang W, Wei S, Li W (2009) Review of pharmacological effects of Glycyrrhiza radix and its bioactive compounds. Zhongguo Zhong Yao Za Zhi 34:2695–2700. doi:10.1002/ptr
55. Agarwal R, Iezhitsa I, Awaludin NA et al (2013) Effects of magnesium taurate on the onset and progression of galactose-induced experimental cataract: in vivo and in vitro evaluation. Exp Eye Res 110:35–43. doi:10.1016/j.exer.2013.02.011
56. Randazzo J, Zhang P, Makita J et al (2011) Orally active multi-functional antioxidants delay cataract formation in streptozotocin (type 1) diabetic and gamma-irradiated rats. PLoS One 6:e18980. doi:10.1371/journal.pone.0018980
57. Sai Varsha MKN, Raman T, Manikandan R (2014) Inhibition of diabetic-cataract by vitamin K1 involves modulation of hyperglycemia-induced alterations to lens calcium homeostasis. Exp Eye Res 128:73–82. doi:10.1016/j.exer.2014.09.007
58. Bloemendal H (1982) Lens proteins. CRC Crit Rev Biochem 12:1–38
59. Graw J (2009) Genetics of crystallins: cataract and beyond. Exp Eye Res 88:173–189. doi:10.1016/j.exer.2008.10.011
60. Groenen PJTA, Merck KB, De Jong WW, Bloemendal H (1994) Structure and modifications of the junior chaperone γ-crystallin—from lens transparency to molecular pathology. Eur J Biochem 225:1–19. doi:10.1111/j.1432-1033.1994.00001.x
61. Sun TX, Liang JJN (1998) Intermolecular exchange and stabilization of recombinant human αA- and αB-crystallin. J Biol Chem 273:286–290. doi:10.1074/jbc.273.1.286
62. Derham BK, Harding JJ (1999) Alpha-crystallin as a molecular chaperone. Prog Retin Eye Res 18:463–509. doi:S1350946298000305
63. Horwitz J (1992) Alpha-crystallin can function as a molecular chaperone. Proc Natl Acad Sci U S A 89:10449–10453. doi:10.1073/pnas.89.21.10449
64. Horwitz J (2003) Alpha-crystallin. Exp Eye Res 76:145–153. doi:10.1016/S0014-4835(02)00278-6
65. Sun Y, MacRae TH (2005) Small heat shock proteins: molecular structure and chaperone function. Cell Mol Life Sci 62:2460–2476. doi:10.1007/s00018-005-5190-4
66. Andley UP (2007) Crystallins in the eye: function and pathology. Prog Retin Eye Res 26:78–98. doi:10.1016/j.preteyeres.2006.10.003
67. Bateman OA, Sarra R, Van Genesen ST et al (2003) The stability of human acidic β-crystallin oligomers and hetero-oligomers. Exp Eye Res 77:409–422. doi:10.1016/S0014-4835(03)00173-8
68. Fu L, Liang JJ-N (2002) Unfolding of human lens recombinant βB2- and γC-crystallins. J Struct Biol 139:191–198. doi:S1047847702005452
69. Berbers GA, Hoekman WA, Bloemendal H et al (1984) Homology between the primary structures of the major bovine beta-crystallin chains. Eur J Biochem 139:467–479. doi:10.1111/j.1432-1033.1984.tb08029.x
70. Lampi KJ, Ma Z, Shih M et al (1997) Sequence analysis of betaA3, betaB3, and betaA4 crystallins completes the identification of the major proteins in young human lens. J Biol Chem 272:2268–2275
71. Downard KM, Kokabu Y, Ikeguchi M, Akashi S (2011) Homology-modelled structure of the βb2B3-crystallin heterodimer studied by ion mobility and radical probe MS. FEBS J 278:4044–4054. doi:10.1111/j.1742-4658.2011.08309.x
72. Van Montfort RLM, Bateman OA, Lubsen NH, Slingsby C (2003) Crystal structure of truncated human betaB1-crystallin. Protein Sci 12:2606–2612. doi:10.1110/ps.03265903.The
73. Dolinska MB, Sergeev YV, Chan MP et al (2009) N-terminal extension of βB1-crystallin: identification of a critical region that modulates protein interaction with βA3-crystallin. Biochemistry 48:9684–9695. doi:10.1021/bi9013984
74. Marín-Vinader L, Onnekink C, Van Genesen ST et al (2006) In vivo heteromer formation. Expression of soluble βA4-crystallin requires coexpression of a heteromeric partner. FEBS J 273:3172–3182. doi:10.1111/j.1742-4658.2006.05326.x

75. Basak A, Bateman O, Slingsby C et al (2003) High-resolution X-ray crystal structures of human γD crystallin (1.25Å) and the R58H mutant (1.15Å) associated with aculeiform cataract. J Mol Biol 328:1137–1147. doi:10.1016/S0022-2836(03)00375-9
76. Kong F, King J (2011) Contributions of aromatic pairs to the folding and stability of long-lived human γD-crystallin. Protein Sci 20:513–518. doi:10.1002/pro.583
77. Slingsby C, Wistow GJ (2014) Functions of crystallins in and out of lens: roles in elongated and post-mitotic cells. Prog Biophys Mol Biol 115:52–67. doi:10.1016/j.pbiomolbio.2014.02.006
78. Hemmingsen JM, Gernert KM, Richardson JS, Richardson DC (1994) The tyrosine corner: a feature of most Greek key beta-barrel proteins. Protein Sci 3:1927–1937. doi:10.1002/pro.5560031104
79. Hamill SJ, Cota E, Chothia C, Clarke J (2000) Conservation of folding and stability within a protein family: the tyrosine corner as an evolutionary cul-de-sac. J Mol Biol 295:641–649. doi:10.1006/jmbi.1999.3360
80. Spector A, Roy D (1978) Disulfide-linked high molecular weight protein associated with human cataract. Proc Natl Acad Sci U S A 75:3244–3248
81. Chen J, Flaugh SL, Callis PR, King J (2006) Mechanism of the highly efficient quenching of tryptophan fluorescence in human γD-crystallin. Biochemistry 45:11552–11563. doi:10.1021/bi060988v
82. Chen J, Toptygin D, Brand L, King JA (2008) Mechanism of the efficient tryptophan fluorescence quenching in human gammaD-crystallin studied by time-resolved fluorescence. Biochemistry 47(40):10705–10721
83. Searle BC, Dasari S, Wilmarth PA et al (2005) Identification of protein modifications using MS/MS de novo sequencing and the OpenSea alignment algorithm. J Proteome Res 4:546–554. doi:10.1021/pr049781j
84. Wilmarth PA, Tanner S, Dasari S et al (2006) Age-related changes in human crystalline determined from comparative analysis of post-translational modifications in young and aged lens: does deamidation contribute to crystallin insolubility? J Proteome Res 5:2554–2566. doi:10.1021/pr050473a
85. MacCoss MJ, McDonald WH, Saraf A et al (2002) Shotgun identification of protein modifications from protein complexes and lens tissue. Proc Natl Acad Sci U S A 99:7900–7905. doi:10.1073/pnas.122231399
86. Kosinski-Collins MS, King JA (2003) In vitro unfolding, refolding, and polymerization of human gammaD crystallin, a protein involved in cataract formation. Protein Sci 12:480–490. doi:10.1110/ps.0225503
87. Kosinski-collins MS, Flaugh SL (2004) Probing folding and fluorescence quenching in human γD crystallin Greek key domains using triple tryptophan mutant proteins. Protein Sci 13(8):2223–2235. doi:10.1110/ps.04627004
88. Flaugh SL, Kosinski-Collins MS, King JA (2005) Interdomain side-chain interactions in human gammaD crystallin influencing folding and stability. Protein Sci 14:2030–2043. doi:10.1110/ps.051460505
89. Flaugh SL, Mills IA, King J (2006) Glutamine deamidation destabilizes human γD-crystallin and lowers the kinetic barrier to unfolding. J Biol Chem 281:30782–30793. doi:10.1074/jbc.M603882200
90. Talla V, Srinivasan N, Balasubramanian D (2008) Visualization of in situ intracellular aggregation of two cataract-associated human γ-crystallin mutants: lose a tail, lose transparency. Invest Ophthalmol Vis Sci 49:3483–3490. doi:10.1167/iovs.07-1114
91. Vendra VPR, Balasubramanian D (2010) Structural and aggregation behavior of the human γD-crystallin mutant E107A, associated with congenital nuclear cataract. Mol Vis 16:2822–2828
92. Schafheimer N, King J (2013) Tryptophan cluster protects human γD-crystallin from ultraviolet radiation-induced photoaggregation in vitro. Photochem Photobiol 89:1106–1115. doi:10.1111/php.12096
93. Moran SD, Zhang TO, Decatur SM, Zanni MT (2013) Amyloid fiber formation in human γD-crystallin induced by UV-B photodamage. Biochemistry 52(36):6169–6181

94. Ji F, Jung J, Koharudin LMI, Gronenborn AM (2013) The human W42R γD-crystallin mutant structure provides a link between congenital and age-related cataracts. J Biol Chem 288:99–109. doi:10.1074/jbc.M112.416354
95. Wu JW, Chen M, Wen W et al (2014) Comparative analysis of human γD-crystallin aggregation under physiological and low pH conditions. PLoS One 9(11):e112309. doi:10.1371/journal.pone.0112309
96. Das P, King JA, Zhou R (2011) Aggregation of γ-crystallins associated with human cataracts via domain swapping at the C-terminal β-strands. Proc Natl Acad Sci U S A 108:10514–10519. doi:10.1073/pnas.1019152108
97. Das P, King JA, Zhou R (2010) β-strand interactions at the domain interface critical for the stability of human lens γD-crystallin. Protein Sci 19:131–140. doi:10.1002/pro.296
98. Shin KHH, Lim SSS, Kim DKK (1998) Effect of byakangelicin, an aldose reductase inhibitor, on galactosemic cataracts, the polyol contents and Na+, K+ATPase activity in sciatic nerves of streptozotocin-induced diabetic rats. Phytomedicine 5:121–127. doi:10.1016/S0944-7113(98)80008-1
99. Nagai R, Nagai M, Shimasaki S et al (2010) Citric acid inhibits development of cataracts, proteinuria and ketosis in streptozotocin (type 1) diabetic rats. Biochem Biophys Res Commun 393:118–122. doi:10.1016/j.bbrc.2010.01.095
100. Swamy MS, Abraham EC (1989) Inhibition of lens crystallin glycation and high molecular weight aggregate formation by aspirin in vitro and in vivo. Invest Ophthalmol Vis Sci 30:1120–1126
101. Shi Q, Yan H, Li M, Harding JJ (2009) Effect of a combination of carnosine and aspirin eye drops on streptozotocin-induced diabetic cataract in rats. Mol Vis 15:2129–2138
102. Gacche RN, Dhole NA (2011) Aldose reductase inhibitory, anti-cataract and antioxidant potential of selected medicinal plants from the Marathwada region, India. Nat Prod Res 25:760–763. doi:10.1080/14786419.2010.536951
103. Dongare V, Kulkarni C, Kondawar M et al (2012) Inhibition of aldose reductase and anti-cataract action of trans-anethole isolated from Foeniculum vulgare Mill. fruits. Food Chem 132:385–390. doi:10.1016/j.foodchem.2011.11.005
104. Rathi SS, Grover JK, Vikrant V, Biswas NR (2002) Prevention of experimental diabetic cataract by Indian ayurvedic plant extracts. Phytother Res 777:774–777
105. Pastene E, Avello M, Letelier ME et al (2007) Preliminary studies on antioxidant and anti-cataract activities of Cheilanthes glauca (Cav.) Mett. through various in vitro models. Electron J Food Plants Chem 2:1–8
106. Vibin M, Siva Priya SG, N Rooban B et al (2010) Broccoli regulates protein alterations and cataractogenesis in selenite models. Curr Eye Res 35:99–107. doi:10.3109/02713680903428991
107. Suryanarayana P, Petrash JM (2004) Inhibition of aldose reductase by tannoid principles of Emblica officinalis: implications for the prevention of sugar cataract. Mol Vis 10:148–154
108. Shin KH, Chung MS, Cho TS (1994) Effects of Furanocoumarins from Angelica dahurica on aldose reductase and galactosemic cataract formation in rats. Arch Pharm Res 5:331–336
109. Lija Y, Biju PG, Reeni A et al (2006) Modulation of selenite cataract by the flavonoid fraction of Emilia sonchifolia in experimental animal models. Phytother Res 20(12):1091–1095. doi:10.1002/ptr
110. Halder N, Joshi S, Gupta SK (2003) Lens aldose reductase inhibiting potential of some indigenous plants. J Ethnopharmacol 86:113–116. doi:10.1016/S0378-8741(03)00052-7
111. Devi VG, Rooban BN, Sasikala V et al (2009) Toxicology in vitro Isorhamnetin-3-glucoside alleviates oxidative stress and opacification in selenite cataract in vitro. Toxicol In Vitro 24:1662–1669. doi:10.1016/j.tiv.2010.05.021
112. Ertekin MV, Koçer I, Karslioğlu I et al (2004) Effects of oral Ginkgo biloba supplementation on cataract formation and oxidative stress occurring in lenses of rats exposed to total cranium radiotherapy. Jpn J Ophthalmol 48:499–502. doi:10.1007/s10384-004-0101-z
113. Vit P, Jacob TJ (2008) Putative anticataract properties of honey studied by the action of flavonoids on a lens culture model. J Heat Sci 54:196–202. doi:10.1248/jhs.54.196

114. Goulet DR, Knee KM, King JA (2011) Inhibition of unfolding and aggregation of lens protein human gamma D crystallin by sodium citrate. Exp Eye Res 93:371–381. doi:10.1016/j.exer.2011.04.011
115. Gong B, Zhang L, Lam DS et al (2010) Sodium 4-phenylbutyrate ameliorates the effects of cataract-causing mutant gammaD-crystallin in cultured cells. Mol Vis 16:997–1003
116. Friberg G, Pande J, Ogun O, Benedek GB (1996) Pantethine inhibits the formation of high-Tc protein aggregates in gamma B crystallin solutions. Curr Eye Res 15:1182–1190
117. Wu JW-RR, Kao C-YY, Lin LT-WW et al (2013) Human γD-crystallin aggregation induced by ultraviolet C irradiation is suppressed by resveratrol. Biochem Eng J 78:189–197. doi:10.1016/j.bej.2013.03.014
118. Liao J, Lin I, Huang K et al (2014) Carnosine ameliorates lens protein turbidity formations by inhibiting calpain proteolysis and ultraviolet C—induced degradation. J Agric Food Chem 62:5932–5938. doi:10.1021/jf5017708
119. Liem-The KN, Stols AL, Jap PH, Hoenders HJ (1975) X-ray induced cataract in rabbit lens. Exp Eye Res 20:317–328. doi:10.1016/0014-4835(75)90114-1

Protein Motions, Dynamic Effects and Thermal Stability in Dihydrofolate Reductase from the Hyperthermophile *Thermotoga maritima*

Rudolf K. Allemann, E. Joel Loveridge, and Louis Y.P. Luk

Abstract Dihydrofolate reductase (DHFR) has long been used as a model system in studies of the relationship between enzyme structure and catalysis. DHFR from the hyperthermophilic bacterium *Thermotoga maritima* (TmDHFR) is substantially different to other chromosomal DHFRs. It is dimeric where most others are monomeric, it lacks the conformational behaviour of monomeric DHFRs, and the kinetics of the catalysed reaction are significantly different. Experimental and computational studies of TmDHFR and comparison to other DHFRs have yielded deep insights into the role of enzyme motions and dynamics in catalysis. Mutational studies and formation of hybrids between TmDHFR and a monomeric homologue have demonstrated that dimerisation is required for extreme thermostability, but also leads to an inability to adequately close the active site with detrimental effects for the speed of the catalysed reaction. However, in common with other DHFRs there is no involvement of large-scale enzyme motions in the chemical reaction itself and dynamic coupling to the reaction coordinate is efficiently minimised. Studies of DHFRs from hyperthermophilic organisms and comparisons to their mesophilic counterparts remain a rich source of information on the fundamental nature of enzyme catalysis.

Keywords Dihydrofolate reductase • Dynamics • Enzyme function • Protein stability • Adaptation

Introduction

The study of protein dynamics is a multidisciplinary field of research, drawing on techniques from genomics, biophysics, biological chemistry and computational modelling. This field has gained much attention from researchers, because the dynamic nature of proteins is a key factor in their function. Dynamic profiles of

R.K. Allemann (✉) • E.J. Loveridge • L.Y.P. Luk
School of Chemistry, Cardiff University,
Main Building, Park Place, Cardiff CF10 3AT, UK
e-mail: AllemannRK@cardiff.ac.uk; loveridgeej@cardiff.ac.uk; Lukly@cardiff.ac.uk

L. Olivares-Quiroz et al. (eds.), *Physical Biology of Proteins and Peptides*,
DOI 10.1007/978-3-319-21687-4_6

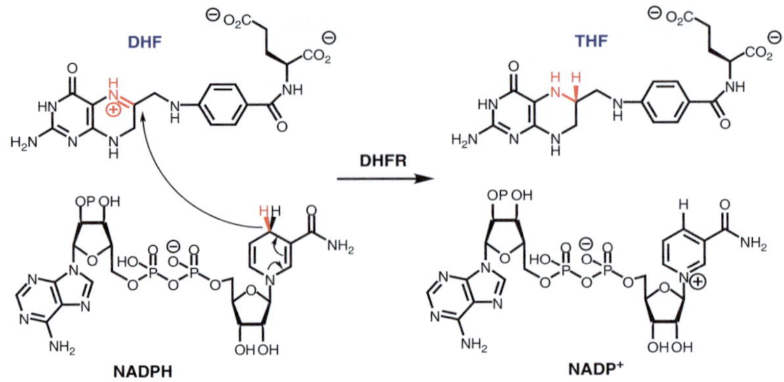

Fig. 1 Reaction catalysed by dihydrofolate reductase

proteins are generally broad, having atomic bond vibrations on the femtosecond timescale, structural adaptation on the millisecond timescale and other fluctuations on timescales in between. It is well established that conformational flexibility of many enzymes is intimately linked to the thermodynamic favourability of protein-ligand interactions [1–4]. Moreover, it has been postulated that enzymes have evolved to use non-equilibrium fluctuations to promote chemical transformations [5–11]. The existence of such 'promoting motions', proposed to modulate the width and/or height of the potential energy barrier to the chemical transformation, has however been vigorously questioned. Indeed, a number of studies have indicated that protein motions are predominantly orthogonal to the chemical coordinate and so have little effect on it [12–18].

Enzymes that catalyse hydrogen transfer reactions (H^+, $H^\bullet$, H^-) have been extensively used in investigations of the effects of protein dynamics on catalysis since such reactions often proceed with a significant contribution from quantum mechanical tunnelling, which is sensitive to changes in the distance between the reacting groups and hence to environmental motions. Dihydrofolate reductase (DHFR) catalyses the reduction of 7,8-dihydrofolate (DHF) to (*S*)-5,6,7,8-tetrahydrofolate (THF) by transferring the C-4 pro-*R* hydrogen from NADPH to the C-6 position of DHF as well as a proton from solvent to the N-5 position (Fig. 1) [18–36], which serves as an one-carbon carrier during the biosynthesis of metabolites such as thymidylate, purines and certain amino acids [37]. Due to its importance for cell survival, DHFR is found in numerous organisms, and homologues isolated from a number of different species have been characterised. It has been suggested that during catalysis by DHFR from *Escherichia coli* (EcDHFR) the chemical step of hydride transfer is enhanced by coupling to non-statistical 'promoting motions' [20, 21, 23, 24, 29, 30]. While direct evidence that supports the existence of these protein fluctuations remains unavailable, a series of experimental and computational investigations have provided convincing data that EcDHFR adopts a relatively static environment during the chemical transformation [14, 22, 27, 31–34, 36]. Indeed, it appears that dynamic coupling is minimised in the transition state species in

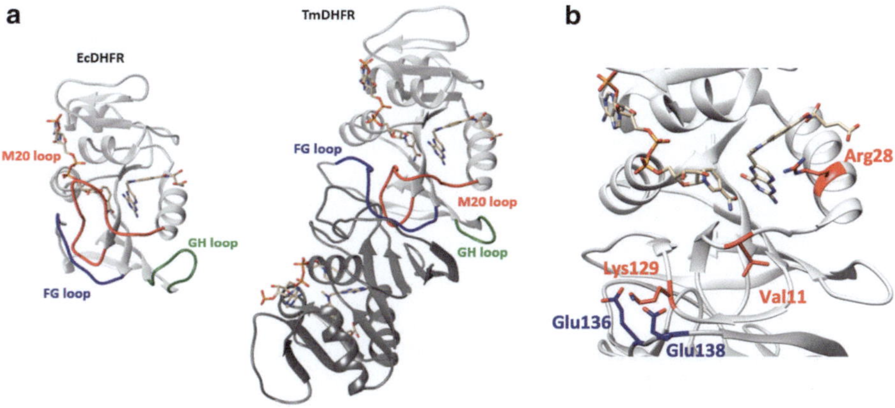

Fig. 2 (**a**) Cartoon representations of EcDHFR (1RX3) [41] and TmDHFR (1D1G) [40]. NADP$^+$ and methrotrexate serve as ligands. (**b**) The active site and interface region of TmDHFR, showing residues in TmDHFR that play critical roles in substrate binding and inter-subunit interactions, including Val11, Arg28, Lys29 in one subunit (*red*) and Glu136 and Glue138 the other subunit (*blue*)

EcDHFR catalysis, reducing the number of non-productive reaction trajectories on the transition state dividing surface [33].

Most chromosomal DHFRs are monomeric and share highly similar tertiary structures. However, DHFR from the hyperthermophilic bacterium *Thermotoga maritima* relies on the formation of a homodimer for stability [38–40] (Fig. 2a). Indeed, with a thermal melting temperature, T_m, of 83 °C, TmDHFR is the most thermostable DHFR characterised thus far [40]. Its stability is achieved by a combination of the dimeric structure with the shortening of surface loops and a reduction in the number of thermolabile residues [39, 40]. However, the adaptations required for thermostability leads to a reduction of the conformational flexibility of TmDHFR. In EcDHFR, there are several mobile segments including the M20 (residues 9–24), FG (residues 116–132) and GH (residues 142–149) loops (Fig. 2a) that coordinate with one another and switch the enzyme between the closed and occluded conformations during the catalytic cycle [1, 41, 42]. This conformational change is critical for the progression through the catalytic cycle, as it facilitates the physical steps of substrate association and product dissociation [1, 41, 42]. This conformational switch is not found in all DHFRs, and is not observed in some DHFRs even though they have similar catalytic cycles to EcDHFR [43–46]. On the other hand, many DHFRs can adopt the closed conformation, which is viewed as the catalytically competent, reaction ready configuration [17, 41, 45, 47]. In TmDHFR the dimeric structure buries the FG loop in the interface, preventing the conformational flexibility observed in EcDHFR and locking the enzyme in an open conformation even with ligands bound and it is believed that no closed conformation can form [40].

The dimeric structure and resulting open active site configuration have profound implications for TmDHFR catalysis. Indeed, hydride transfer catalysed by TmDHFR has a much lower rate constant than for other chromosomal DHFRs [42, 48] and catalytic turnover at pH 7 is partially limited by hydride transfer [48]. Furthermore, the primary kinetic isotope effect (KIE) measured with [4*R*-^{2}H]-NADPH under single turnover conditions is mostly temperature independent above 25 °C but temperature dependent below this point, thus showing biphasic behaviour [49]. Since the quaternary structure, conformational flexibility and kinetics of the catalysed reaction for TmDHFR are significantly different from those of EcDHFR, the investigation of TmDHFR provides significant insights into the role of enzyme motions and dynamics in catalysis.

Dimerisation Is Essential for Thermostability But Affects the Catalytic Ability of TmDHFR

The extreme thermal stability of TmDHFR appears to be mostly due to the dimeric nature of this enzyme [40, 50]. Furthermore, the role of the loop regions in TmDHFR and EcDHFR is markedly different (*vide supra*). Introducing residues from the dimer interface of TmDHFR into the equivalent region of EcDHFR led to a dimeric EcDHFR-TmDHFR hybrid [51]. Interestingly, such a hybrid shares similar single turnover and steady state rate constants with wild type TmDHFR under physiological conditions, despite retaining 94 % of the original EcDHFR sequence [51]. Hence dimerisation of EcDHFR appears to have a detrimental effect on the geometry and electrostatics of the active site, although structural studies would be required to determine the exact effect. The hybrid did however retain EcDHFR-like substrate binding properties, as the ligand-contacting residues were unaltered, and the temperature dependence of the KIE was similar to that of EcDHFR [51]. The dimeric hybrid is considerably more thermostable than EcDHFR, but not as thermostable as TmDHFR, most probably because the thermal stability of the hybrid is limited by unfolding of other regions of the enzyme such as the hinge region around Asp87 of EcDHFR that are afforded no additional protection by dimerization [51, 52].

In contrast, although a monomeric TmDHFR-EcDHFR hybrid could not be constructed using similar methodology to formation of the dimeric EcDHFR-TmDHFR hybrid [51], disruption of the TmDHFR dimer interface by incorporation of acidic residues causes a large drop in thermal stability regardless of whether dimerisation itself is prevented [50]. Monomerisation does not have a large effect on the single turnover rate constants, although it does compromise the steady state rate constant significantly [50]. Interestingly, the TmDHFR-V11D variant has a monophasic temperature dependence of the KIE in both the monomeric and dimeric state (Figs. 2b and 3) [50]. Removal of an inter-subunit salt bridge between K129 of one subunit and E136 and E138 of the other, which also reduces the thermal stability but does not cause monomerisation, leads to monophasic kinetic behaviour (Fig. 2b) [53]. It therefore appears that perturbation of the dimer interface leads to changes in the active site that affect the reaction kinetics.

As expected, dimerisation of EcDHFR appears to restrain the catalytically important loops by creating inter-subunit interactions that disfavour normal movement of those loops and decreases the rate of catalytic turnover. In the case of

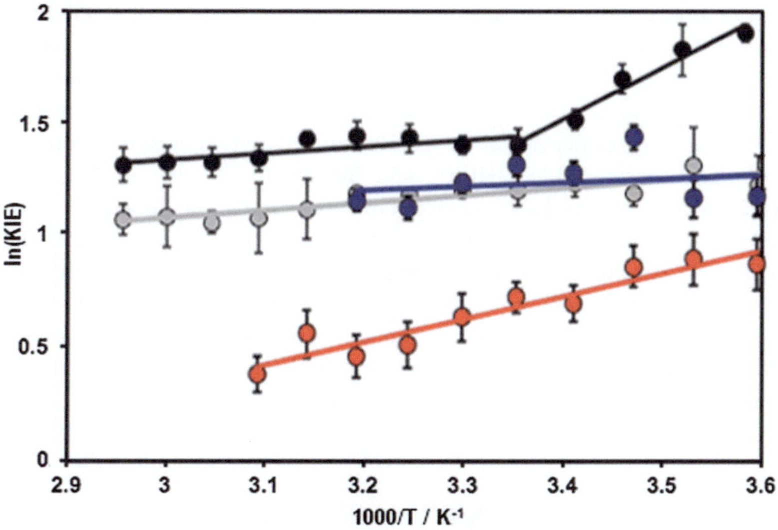

Fig. 3 The temperature dependence of the KIE on hydride transfer for TmDHFR (*black circle*), TmDHFR-K129E (*grey circle*), dimeric TmDHFR-V11D (*blue circle*) and monomeric TmDHFR-V11D (*red circle*). The natural logarithm of the KIE is plotted against the inverse temperature

TmDHFR monomerisation, the increase in the flexibility of the loop regions normally involved in dimerisation [54] is detrimental to catalytic turnover but not to hydride transfer [50] providing further evidence that hydride transfer is uncoupled from thermal motions of the enzyme. A similar network of coupled motions to that seen in EcDHFR [20, 21] has been identified in TmDHFR, which spans the dimer interface [55]. It was suggested that the hypothetical monomer of TmDHFR would have a reduced correlation of motions relative to the native dimer [55], although the loop regions would have greater flexibility generally [54].

The major consequence of the altered loops of TmDHFR is that the enzyme is locked in an open conformation, indicating that conformational sampling must be subject to high energy barriers with a much more solvent-exposed active site than in EcDHFR [40, 41] and a dramatic change in the active site electrostatics. Replacement of the central four residues (Met16-Ala19) of the M20 loop of EcDHFR with a single glycine, likely increasing the solvent accessibility of the active site, leads to similar single turnover rate constants at pH 7 to TmDHFR [56], and hydride transfer is partially rate limiting in the steady state as seen for TmDHFR [48]. Interestingly, the presence of an arginine residue (R28) within the active site of TmDHFR, which would strongly affect the active site electrostatics, was found to have little effect on the protonation of dihydrofolate or on the single turnover rate constant, but instead was responsible for tight binding of the substrate (Fig. 2b) [57].

TmDHFR therefore appears to have been adapted for thermostability at the expense of catalytic speed. Dimerisation is required for extreme thermostability, but the modifications to the enzyme that allow dimerisation also appear to have rendered the loop regions unable to perform a similar function to those of EcDHFR. In the monomeric thermophilic DHFR from *Geobacillus stearothermophilus* (BsDHFR), whose host organism grows optimally at 30–60 °C, the removal of thermolabile residues and the extension of secondary structural elements contribute to thermal adaptation [58, 59]. All hyperthermostable DHFRs may have evolved through dimerisation of monomeric thermostable DHFRs, and the results for the dimeric EcDHFR-TmDHFR hybrid described above suggest that, once the initial dimerisation had constrained the loops regions and impaired catalysis, additional alterations improved thermostability but prevented the loop regions from reverting to their original flexibility. As discussed elsewhere [51, 57], TmDHFR likely has a reduced requirement for 'high' DHFR activity due to the presence of a flavin-dependent thymidylate synthase that regenerates THF rather than forming DHF, and the predominant evolutionary pressure is therefore for thermostability rather than for catalysis.

Long Range Motions Do Not Couple to the Chemical Step of TmDHFR Catalysis

Equilibrium motions of an enzyme can be altered by the addition of organic co-solvents, which alter the viscosity of the medium, typically decrease its dielectric constant and affect the surface hydration of the enzyme. In all cases, co-solvents alter the conformational coordinate of enzyme catalysis [60, 61]. Whereas both surface and interior protein motions are slowed by increasing solvent viscosity [62], only exterior motions are affected by decreasing dielectric constant of the medium [63]. Protein motions coupled to that of the hydration water layer could also be affected by co-solvent addition [64]. This allows the effect of long-range protein motions on the chemical step of the catalytic cycle to be probed by changing the solvent composition.

We investigated the kinetic behaviour of TmDHFR in water-cosolvent mixtures with dielectric constants of 40–80 and viscosities of 1–8 mPa s (at 20 °C; equivalent values were 37–73 and 0.7–4 mPa s respectively at 40 °C). Under steady-state conditions TmDHFR and EcDHFR behave differently in the presence of organic co-solvents; whereas dielectric constant caused no apparent effect [65], viscosity alters the steady state turnover rate constants for EcDHFR but not TmDHFR [13]. These observations are consistent with the differing conformational behaviour of these enzymes, considering that there is a substantial conformational change involved in the catalytic turnover of EcDHFR but structural changes in the catalytic cycle of TmDHFR are negligible. Decreasing the dielectric constant, but not the viscosity of the medium, reduces the single turnover hydride transfer rate constants of both enzymes. Nevertheless, the single turnover KIE at pH 7 is insensitive to changes of solvent composition. Similar observations were made for the moderately thermophilic BsDHFR [59] and the cold-adapted homologue isolated from *Moritella profunda* (MpDHFR) [16].

Most interestingly, although the magnitude of the KIE is not significantly affected by solvent composition, differences in the temperature dependence of the KIE are

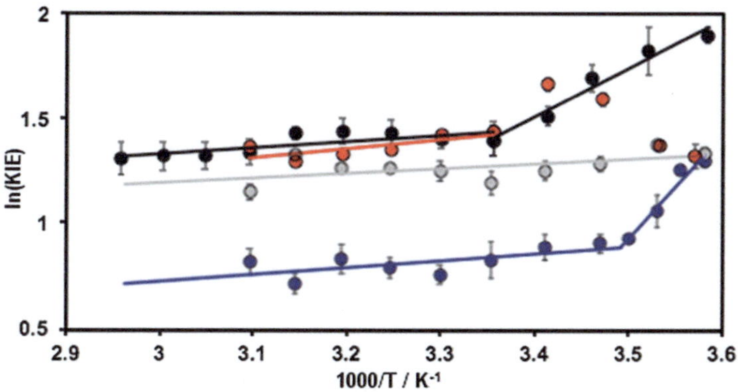

Fig. 4 The temperature dependence of the KIE on hydride transfer for TmDHFR in buffer alone (*black circle*), and in buffer containing 30 % sucrose (*grey circle*), 50 % glycerol (*blue circle*) and 50 % methanol (*red circle*). The natural logarithm of the KIE is plotted against the inverse temperature

observed in certain solvents [65] (Fig. 4). Methanol causes no change in the KIE above 25 °C, but leads to increased kinetic complexity in the hydride transfer step below this temperature, causing a decrease in the KIE. However, increasing concentrations of glycerol lead to a decrease in the magnitude of the KIE and increase the range of the temperature-independent regions of the KIE, whereas sucrose causes the KIE to become entirely temperature independent over the temperature range studied. It is therefore likely that glycerol and sucrose induce changes to the interactions in the dimer interface similar to those seen in TmDHFR variants with altered dimer interfaces (*vide supra*).

However, while co-solvents clearly induce long-range effects on TmDHFR catalysis, most likely by interfering with the inter-subunit interactions, such effects do not imply a direct role for long-range TmDHFR motions in the catalysed hydride transfer. Indeed, the results for all four DHFRs together indicate that long-range motions are unlikely to play a role in the chemical coordinate, since viscosity did not affect the single turnover rate constants [13, 16, 59, 65]. Instead, the addition of organic co-solvents likely perturbs the conformational ensemble of the enzyme, altering the pre-organisation process required to achieve an electrostatically favourable environment for hydride transfer [13, 16, 59]. This would lead to a rise in the reorganisation energy, and a consequent reduction in the hydride transfer rate constant. However, studies with organic co-solvents do not rule out a short range, local 'promoting motion'.

Local Motions Do Not Couple to the Chemical Step of TmDHFR Catalysis

Recently, 'heavy' enzyme experiments have been created to investigate the effect of enzyme dynamics on catalysis [32, 33, 36, 66–72]. In these studies, non-exchangeable C, N and H positions of the enzyme are substituted with their

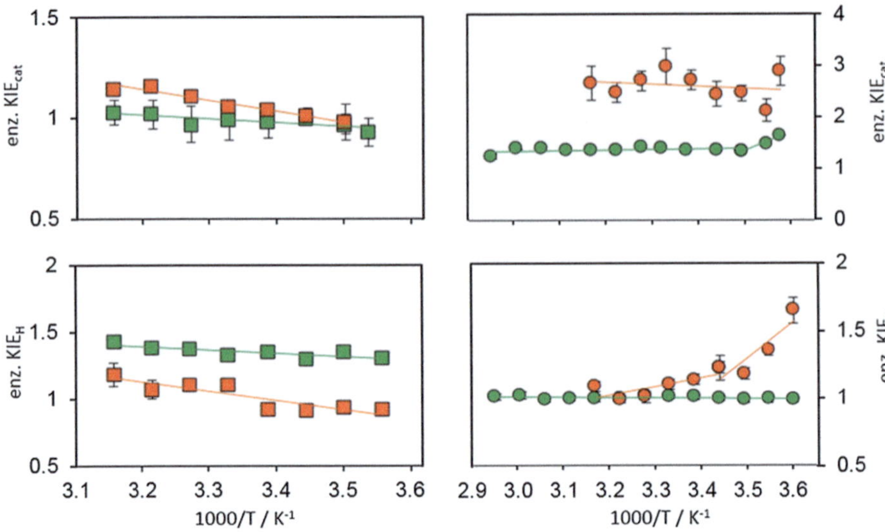

Fig. 5 Enzyme KIEs at pH 7 for EcDHFR (*orange square*) [32], EcDHFR-N23PP/S148A (*green square*) [33], TmDHFR (*green circle*) [71] and BsDHFR (*orange circle*) [72] under steady state (*top*) and pre-steady state (*bottom*) conditions

corresponding heavy isotopes ^{13}C, ^{15}N and ^{2}H. This increases the molecular weight of the enzyme by ~10 % relative to the corresponding 'light' enzyme with isotopes of natural abundance. As mass-dependent atomic vibrations are sensitive to heavy isotope substitution, the entire dynamic profile of the enzyme is consequently slowed. In contrast, within the Born-Oppenheimer approximation the electrostatics of the enzyme remain unaffected by heavy isotope substitution [66, 67]. Kinetic investigations of isotopically labelled enzymes therefore reveal information about the nature of enzyme dynamics. HIV-1 protease, purine nucleoside phosphorylase, alanine racemase, pentaerythritol tetranitrate reductase and DHFR have all been subjected to this approach [32, 33, 36, 66–72]. In most cases, the chemical transformations catalysed by the 'heavy' enzymes had slightly lower rate constants than those of the corresponding 'light' enzymes.

'Heavy' enzyme analyses have been performed on four DHFRs that have distinct catalytic behaviour and biophysical properties, EcDHFR, EcDHFR-N23PP/S148A, BsDHFR and TmDHFR (Fig. 5) [32, 33, 71, 72]. Under steady-state conditions, protein isotope labelling of EcDHFR caused a slight temperature-dependent increase of enzyme KIE_{cat} ($k_{cat}{}^{LE}/k_{cat}{}^{HE}$, where LE and HE indicate 'light' and 'heavy' enzyme, respectively) [32]. In contrast, for the mutant EcDHFR-N23PP/S148A, which cannot adopt the occluded conformation, the enzyme KIE_{cat} remains ~1 at all examined temperatures [33]. In the case of BsDHFR, the most flexible homologue among these enzymes [72–74], the enzyme KIE_{cat} was large (~2.6) but temperature-independent [72]. Because the physical steps in these enzymes predominantly limited catalytic turnover, these observations implied that the magnitude of the enzyme

KIE_{cat} was affected by the intrinsic flexibility of the enzyme, while its temperature dependence was caused by the presence of a conformational change coupled to the rate-limiting physical step of catalysis.

For TmDHFR, the enzyme KIE_{cat} was ~1.35 above 15 °C but increased sharply below this temperature, reaching 1.7 at 5 °C [71] (Fig. 5). Interestingly, steady-state kinetics measured with TmDHFR also revealed a subtle biphasic behaviour, which disappeared when the dimer interface was disrupted [50, 53]. These results together suggest that the proposed conformational change stems from alterations to the inter-subunit interactions at low temperatures [71].

The enzyme KIE_H measured under pre-steady-state conditions (k_H^{LE}/k_H^{HE}) also varied greatly among these enzymes (Fig. 5) [32, 33, 71, 72]. For EcDHFR there was a steady, temperature-dependent increase in the enzyme KIE_H [32], but the magnitude of this parameter was greatly enhanced in EcDHFR-N23PP/S148A [33]. Interestingly, the temperature dependence of the enzyme KIE_H measured for BsDHFR was reversed and noticeably stronger. While the enzyme KIE_H was small in the physiological range, it increased sharply below 15 °C [72]. Based on computational analyses, the observed kinetic differences are indications of dynamic coupling to the chemical coordinate [32, 33, 72]. However, such a coupling does not modulate the shape of the barrier and promote hydride transfer. Instead, the coupling of environmental motions to the chemical coordinate induces additional dynamic recrossing, increasing the number of non-productive reaction trajectories on the transition state dividing surface that allow the product to revert to the substrate side of the transition state dividing surface. As such, dynamic coupling to the chemical coordinate in DHFR appears to be detrimental to the catalysed reaction.

Unlike with other DHFRs, protein isotope labelling exerted no effect on the single turnover rate constants for TmDHFR-catalysed hydride transfer giving an enzyme KIE_H of ~1 at all temperatures (Fig. 5) [71]. Accordingly, mass-dependent protein dynamics do not cause any detectable effect on the chemical coordinate. Therefore, local motions do not couple to the chemical coordinate in TmDHFR catalysis. As dynamic coupling is detrimental to DHFR catalysis, it appears that TmDHFR has gained a slight benefit by eliminating these dynamic recrossing trajectories [71, 72]. However, this beneficial effect is negligible compared to the overall energy barrier to hydride transfer. Because of the inter-subunit interactions required for dimer formation and the resulting thermostability [39, 40], the conformational sampling required to form a closed active site conformation is likely subject to such an exceptionally high enthalpy of activation that, even though part of this barrier is compensated by the enzyme's physiological temperature (~80 °C), hydride transfer activity remains noticeably lower than its monomeric counterparts [71, 72]. Our site-directed mutagenesis experiments, solvent effect and enzyme isotope labelling investigations have revealed that, while the inter-subunit interactions of TmDHFR are important for the physical steps during catalysis, the enzyme chemistry is decoupled from any protein fluctuations, including long-range slow motions and short-range fast vibrations [50, 53, 57, 65, 71]. These findings are in line with observations made for other DHFRs where dynamic effects are generally minimised under physiological conditions [18].

DHFRs from Other *Thermotoga* Species Are Also Likely to Be Dimeric

Although TmDHFR is the only chromosomal DHFR, for which direct experimental and structural evidence for dimerisation exists [38, 40, 50, 51], sequences of the DHFRS of a number of other *Thermotoga* species are available. These are all highly similar to one another (Fig. 6). Indeed, the residues that form the core of the dimer interface in TmDHFR are almost identical in all five enzymes, with only conservative variations. The residues on the periphery of the dimer interface are less well conserved, but the three-membered salt bridge shown to be important for thermostability [53] is seen in all cases, even if the precise residues are altered. It is also interesting to note that all five *Thermotoga* DHFRs possess the active site R28 residue important for tight binding of the substrate [50]. After *Thermotoga neapolitana* DHFR, which shows 76 % identity to TmDHFR, the next closest match (from NCBI BLAST) is from *Mesoaciditoga lauensis*, with only 46 % identity to TmDHFR.

The high level of identity between the five *Thermotoga* DHFRs, particularly in the dimer interface region, suggests that all are dimeric and that all will show similar thermostability to TmDHFR. The exact thermostability for each of the enzymes is likely to provide valuable information on the adaptations required for DHFR function at extreme temperatures. Furthermore, the high level of identity suggests that their structures will also be similar and that all will be unable to adopt a closed active site conformation and have similarly impaired catalytic activity when compared to less thermostable, monomeric DHFRs. However, many other hyperthermophilic organisms do not possess a DHFR at all [75], complicating attempts to make detailed comparisons. DHFR sequences from hyperthermophilic organisms other than those in the genus *Thermotoga* are required before generalisations about the requirements for extreme thermostability, and its implications for catalysis, can be made.

Conclusion

DHFR from the hyperthermophile *T. maritima* is substantially different to other chromosomal DHFRs studied to date in terms of its quaternary structure and conformational behaviour, which in turn cause important differences in the kinetics of the catalysed reaction. Experimental and computational investigations of the differences, and more importantly the similarities, between TmDHFR and other DHFRs, have allowed significant insights into the role of enzyme motions and dynamics in catalysis to be gained. While dimerisation is required for extreme thermostability, this has detrimental effects on the enzyme's ability to catalyse the desired hydride transfer reaction. Inability to adequately close the active site leads to a dramatically poorer electrostatic environment for the reaction. However, other aspects of TmDHFR catalysis are remarkably similar to other DHFRs. There is no involvement of large-scale enzyme motions in the chemical reaction itself, and dynamic coupling to the reaction coordinate is efficiently minimised. In fact, it is more efficiently minimised in TmDHFR than in other DHFRs, suggesting that the lack of

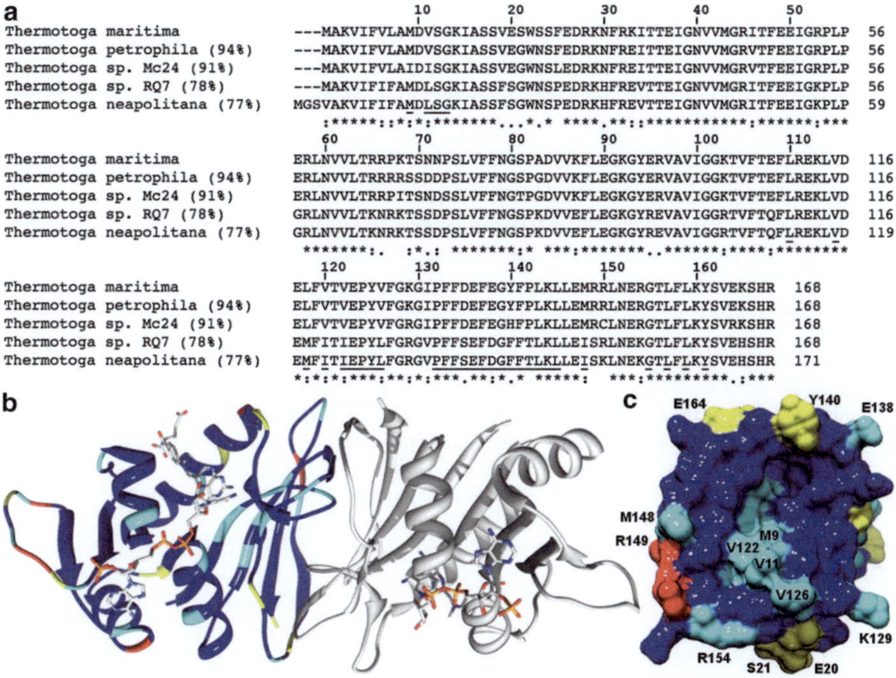

Fig. 6 (**a**) Amino acid alignment of the five sequenced *Thermotoga* DHFRs, indicating the sequence identity to TmDHFR. *Symbols* indicate residues that are fully (*), highly (:) and moderately (.) conserved. Residues involved in the dimer interface are underlined. (**b**) Cartoon representation of TmDHFR and C) surface representation of the dimer interface of TmDHFR indicating the level of sequence similarity within the genus. In panels **b** and **c**, colours indicate residues that are fully (*blue*), highly (*cyan*), moderately (*yellow*) and poorly (*red*) conserved within the genus

flexibility may be beneficial in this regard at least. Studies of TmDHFR and DHFRs from other hyperthermophilic organisms will continue to yield information about enzyme catalysis that studies of the monomeric DHFRs alone may not reveal.

Acknowledgements The authors would like to acknowledge the many contributions made by the members of the Allemann group.

This work was supported by grants BB/L020394/1 and BB/J005266/1 from the UK Biotechnology and Biological Sciences Research Council (BBSRC) and EP/L027240/1 from the UK Engineering and Physical Sciences Research Council (EPSRC). We acknowledge the continuing support by Cardiff University.

References

1. Boehr DD, McElheny D, Dyson HJ, Wright PE (2006) The dynamic energy landscape of dihydrofolate reductase catalysis. Science 313(5793):1638–1642
2. Henzler-Wildman K, Lei M, Thai V, Kerns SJ, Karplus M, Kern D (2007) A hierarchy of timescales in protein dynamics is linked to enzyme catalysis. Nature 450(7171):913

3. Weikl TR, Boehr DD (2012) Conformational selection and induced changes along the catalytic cycle of *Escherichia coli* dihydrofolate reductase. Proteins Struct Funct Bioinf 80(10): 2369–2383
4. Wysoczanski P, Mart RJ, Loveridge EJ, Williams C, Whittaker SBM, Crump MP, Allemann RK (2012) NMR solution structure of a photoswitchable apoptosis activating Bak peptide bound to Bcl-xL. J Am Chem Soc 134(18):7644–7647
5. Knapp MJ, Klinman JP (2002) Environmentally coupled hydrogen tunneling—linking catalysis to dynamics. Eur J Biochem 269(13):3113–3121
6. Mincer JS, Schwartz SD (2004) Rate-promoting vibrations and coupled hydrogen-electron transfer reactions in the condensed phase: a model for enzyme catalysis. J Chem Phys 120: 7755–7760
7. Antoniou D, Schwartz SD (2006) Protein dynamics and enzymatic chemical barrier passage. Philos Trans R Soc B 361:1433–1438
8. Schwartz SD, Schramm VL (2009) Enzymatic transition states and dynamic motion in barrier crossing. Nat Chem Biol 5(8):552–559. doi:10.1038/nchembio.202
9. Pudney CR, Hay S, Levy C, Pang J, Sutcliffe MJ, Leys D, Scrutton NS (2009) Evidence to support the hypothesis that promoting vibrations enhance the rate of an enzyme catalyzed H-tunneling reaction. J Am Chem Soc 131(47):17072–17073
10. Klinman JP (2009) An integrated model for enzyme catalysis emerges from studies of hydrogen tunneling. Chem Phys Lett 471:179–193
11. Hay S, Scrutton NS (2012) Good vibrations in enzyme-catalysed reactions. Nat Chem 4: 161–168
12. Pisliakov AV, Cao J, Kamerlin SCL, Warshel A (2009) Enzyme millisecond conformational dynamics do not catalyze the chemical step. Proc Natl Acad Sci U S A 106:17359–17364
13. Loveridge EJ, Tey L-H, Allemann RK (2010) Solvent effects on catalysis by *Escherichia coli* dihydrofolate reductase. J Am Chem Soc 132(3):1137–1143
14. Adamczyk AJ, Cao J, Kamerlin SCL, Warshel A (2011) Catalysis by dihydrofolate reductase and other enzymes arises from electrostatic preorganization, not conformational motions. Proc Natl Acad Sci U S A 108:14115–14120
15. Boekelheide N, Salomón-Ferrer R, Miller TF (2011) Dynamics and dissipation in enzyme catalysis. Proc Natl Acad Sci U S A 108(39):16159–16163. doi:10.1073/pnas.1106397108
16. Loveridge EJ, Tey L-H, Behiry EM, Dawson WM, Evans RM, Whittaker SB-M, Gunther UL, Williams C, Crump MP, Allemann RK (2011) The role of large-scale motion in catalysis by dihydrofolate reductase. J Am Chem Soc 133:20561–20570
17. Loveridge EJ, Behiry EM, Guo J, Allemann RK (2012) Evidence that a 'dynamic knockout' in *Escherichia coli* dihydrofolate reductase does not affect the chemical step of catalysis. Nat Chem 4(4):292–297
18. Luk LYP, Loveridge EJ, Allemann RK (2015) Protein motions, dynamic effects and enzyme catalysis. Phys Chem Chem Phys. doi:10.1039/C5CP00794A
19. Cameron CE, Benkovic SJ (1997) Evidence for a functional role of the dynamics of Glycine-121 of *Escherichia coli* dihydrofolate reductase obtained from kinetic analysis of a site-directed mutant. Biochemistry 36:15792–15800
20. Radkiewicz JL, Brooks CL III (2000) Protein dynamics in enzymatic catalysis: exploration of dihydrofolate reductase. J Am Chem Soc 122(2):225–231
21. Agarwal PK, Billeter SR, Rajagopalan PTR, Benkovic SJ, Hammes-Schiffer S (2002) Network of coupled promoting motions in enzyme catalysis. Proc Natl Acad Sci U S A 99(5):2794–2799
22. Garcia-Viloca M, Truhlar DG, Gao J (2003) Reaction-path energetics and kinetics of the hydride transfer reaction catalyzed by dihydrofolate reductase. Biochemistry 42(46):13558–13575
23. Rod TH, Radkiewicz JL, Brooks CL III (2003) Correlated motions and the effect of distal mutation in dihydrofolate reductase catalysis. Proc Natl Acad Sci U S A 100:6980–6985
24. Watney LB, Agarwal PK, Hammes-Schiffer S (2003) Effect of mutation on enzyme motion in dihydrofolate reductase. J Am Chem Soc 125:3745–3750
25. Pu J, Ma S, Gao J, Truhlar DG (2005) Small temperature dependence of the kinetic isotope effect for the hydride transfer reaction catalyzed by *Escherichia coli* dihydrofolate reductase. J Phys Chem B 109(18):8551–8556

26. Wang L, Tharp S, Selzer T, Benkovic SJ, Kohen A (2006) Effects of a distal mutation on active site chemistry. Biochemistry 45(5):1383–1392. doi:10.1021/bi0518242
27. Liu H, Warshel A (2007) Origin of the temperature dependence of isotope effects in enzymatic reactions: the case of dihydrofolate reductase. J Phys Chem B 111(27):7852–7861
28. Loveridge EJ, Allemann RK (2009) Direct methods for the analysis of quantum-mechanical tunnelling: dihydrofolate reductase. In: Quantum tunnelling in enzyme-catalysed reactions. The Royal Society of Chemistry, London, pp 179–198
29. Stojković V, Perissinotti LL, Lee J, Benkovic SJ, Kohen A (2010) The effect of active-site isoleucine to alanine mutation on the DHFR catalyzed H-transfer. Chem Commun 46: 8974–8976
30. Bhabha G, Lee J, Ekiert DC, Gam J, Wilson IA, Dyson HJ, Benkovic SJ, Wright PE (2011) A dynamic knockout reveals that conformational fluctuation influence the chemical step of enzyme catalysis. Science 332:234–238
31. Fan Y, Cembran A, Ma S, Gao J (2013) Connecting protein conformational dynamics with catalytic function as illustrated in dihydrofolate reductase. Biochemistry 52(12):2036–2049
32. Luk LYP, Ruiz-Pernia JJ, Dawson WM, Roca M, Loveridge EJ, Glowacki DR, Harvey JN, Mulholland AJ, Tuñón I, Moliner V, Allemann RK (2013) Unraveling the role of protein dynamics in dihydrofolate reductase catalysis. Proc Natl Acad Sci U S A 110:16344–16349
33. Ruiz-Pernia JJ, Luk LYP, García-Meseguer R, Martí S, Loveridge EJ, Tuñón I, Moliner V, Allemann RK (2013) Increased dynamic effects in a catalytically compromised variant of *Escherichia coli* dihydrofolate reductase. J Am Chem Soc 135(49):18689–18696
34. Roston D, Kohen A, Doron D, Major DT (2014) Simulations of remote mutants of dihydrofolate reductase reveal the nature of a network of residues coupled to hydride transfer. J Comput Chem 35(19):1411
35. Singh P, Sen A, Francis K, Kohen A (2014) Extension and limits of the network of coupled motions correlated to hydride transfer in dihydrofolate reductase. J Am Chem Soc 136(6):2575–2582
36. Wang Z, Singh PN, Czekster CM, Kohen A, Schramm VL (2014) Protein mass-modulated effects in the catalytic mechanism of dihydrofolate reductase: beyond promoting vibrations. J Am Chem Soc 136(23):8333
37. Blakley RL (1984) Folates and pterins. Wiley, New York
38. Dams T, Bohm G, Auerbach G, Bader G, Schuring H, Jaenicke R (1998) Homo-dimeric recombinant dihydrofolate reductase from *Thermotoga maritima* shows extreme intrinsic stability. Biol Chem 379(3):367–371
39. Dams T, Jaenicke R (1999) Stability and folding of dihydrofolate reductase from the hyperthermophilic bacterium *Thermotoga maritima*. Biochemistry 38(28):9169–9178
40. Dams T, Auerbach G, Bader G, Jacob U, Ploom T, Huber R, Jaenicke R (2000) The crystal structure of dihydrofolate reductase from *Thermotoga maritima*: molecular features of thermostability. J Mol Biol 297(3):659–672
41. Sawaya MR, Kraut J (1997) Loop and subdomain movements in the mechanism of *Escherichia coli* dihydrofolate reductase: crystallographic evidence. Biochemistry 36(3): 586–603
42. Fierke CA, Johnson KA, Benkovic SJ (1987) Construction and evaluation of the kinetic scheme associated with dihydrofolate reductase from *Escherichia coli*. Biochemistry 26(13): 4085–4092
43. Andrews J, Fierke CA, Birdsall B, Ostler G, Feeney J, Roberts GCK, Benkovic SJ (1989) A kinetic study of wild-type and mutant dihydrofolate reductases from *Lactobacillus casei*. Biochemistry 28(14):5743–5750. doi:10.1021/bi00440a007
44. Appleman JR, Beard WA, Delcamp TJ, Prendergast NJ, Freisheim JH, Blakley RL (1990) Unusual transient-state and steady-state kinetic behavior is predicted by the kinetic scheme operational for recombinant human dihydrofolate reductase. J Biol Chem 265(5):2740–2748
45. Bhabha G, Ekiert DC, Jennewein M, Zmasek CM, Tuttle LM, Kroon G, Dyson HJ, Godzik A, Wilson IA, Wright PE (2013) Divergent evolution of protein conformational dynamics in dihydrofolate reductase. Nat Struct Mol Biol 20(11):1243–1249

46. Behiry EM, Luk LYP, Matthews SM, Loveridge EJ, Allemann RK (2014) Role of the occluded conformation in bacterial dihydrofolate reductases. Biochemistry 53(29):4761–4768
47. Hay S, Evans RM, Levy C, Loveridge EJ, Wang X, Leys D, Allemann RK, Scrutton NS (2009) Are the catalytic properties of enzymes from piezophilic organisms pressure adapted? Chembiochem 10(14):2348–2353
48. Maglia G, Javed MH, Allemann RK (2003) Hydride transfer during catalysis by dihydrofolate reductase from *Thermotoga maritima*. Biochem J 374:529–535
49. Maglia G, Allemann RK (2003) Evidence for environmentally coupled hydrogen tunneling during dihydrofolate reductase catalysis. J Am Chem Soc 125(44):13372–13373
50. Loveridge EJ, Rodriguez RJ, Swanwick RS, Allemann RK (2009) Effect of dimerization on the stability and catalytic activity of dihydrofolate reductase from the hyperthermophile *Thermotoga maritima*. Biochemistry 48(25):5922–5933
51. Guo J, Loveridge EJ, Luk LYP, Allemann RK (2013) Effect of dimerization on dihydrofolate reductase catalysis. Biochemistry 52(22):3881–3887
52. Tey LH, Loveridge EJ, Swanwick RS, Flitsch SL, Allemann RK (2010) Highly site-selective stability increases by glycosylation of dihydrofolate reductase. FEBS J 277(9):2171–2179
53. Loveridge EJ, Allemann RK (2010) The temperature dependence of the kinetic isotope effects of dihydrofolate reductase from *Thermotoga maritima* is influenced by intersubunit interactions. Biochemistry 49(25):5390–5396
54. Pang J, Allemann RK (2007) Molecular dynamics simulation of thermal unfolding of Thermatoga maritima DHFR. Phys Chem Chem Phys 9(6):711–718
55. Pang JY, Pu JZ, Gao JL, Truhlar DG, Allemann RK (2006) Hydride transfer reaction catalyzed by hyperthermophilic dihydrofolate reductase is dominated by quantum mechanical tunneling and is promoted by both inter- and intramonomeric correlated motions. J Am Chem Soc 128(24):8015–8023
56. Li L, Falzone CJ, Wright PE, Benkovic SJ (1992) Functional role of mobile loop of *Escherichia coli* dihydrofolate reductase in transition-state stabilization. Biochemistry 31:7826–7833
57. Loveridge EJ, Maglia G, Allemann RK (2009) The role of arginine 28 in catalysis by dihydrofolate reductase from the hyperthermophile *Thermotoga maritima*. Chembiochem 10(16): 2624–2627
58. Kim HS, Damo SM, Lee SY, Wemmer D, Klinman JP (2005) Structure and hydride transfer mechanism of a moderate thermophilic dihydrofolate reductase from *Bacillus stearothermophilus* and comparison to its mesophilic and hyperthermophilic homologues. Biochemistry 44(34):11428–11439
59. Guo J, Luk LYP, Loveridge EJ, Allemann RK (2014) Thermal adaptation of dihydrofolate reductase from the moderate thermophile *Geobacillus stearothermophilus*. Biochemistry 53:2855–2863
60. Warshel A (1984) Dynamics of enzymatic reactions. Proc Natl Acad Sci U S A 81(2): 444–448
61. Warshel A, Sharma PK, Kato M, Xiang Y, Liu H, Olsson MHM (2006) Electrostatic basis for enzyme catalysis. Chem Rev 106(8):3210–3235
62. Walser R, van Gunsteren WF (2001) Viscosity dependence of protein dynamics. Proteins Struct Funct Bioinf 42(3):414–421
63. Affleck R, Haynes CA, Clark DS (1992) Solvent dielectric effects on protein dynamics. Proc Natl Acad Sci U S A 89(11):5167–5170
64. Fenimore PW, Frauenfelder H, McMahon BH, Parak FG (2002) Slaving: Solvent fluctuations dominate protein dynamics and functions. Proc Natl Acad Sci U S A 99(25):16047–16051
65. Loveridge EJ, Evans RM, Allemann RK (2008) Solvent effects on environmentally coupled hydrogen tunnelling during catalysis by dihydrofolate reductase from *Thermotoga maritima*. Chem Eur J 14(34):10782–10788
66. Silva RG, Murkin AS, Schramm VL (2011) Femtosecond dynamics coupled to chemical barrier crossing in a Born-Oppenheimer enzyme. Proc Natl Acad Sci U S A 108:18661–18665
67. Kipp DR, Silva RG, Schramm VL (2011) Mass-dependent bond vibrational dynamics influence catalysis by HIV-1 protease. J Am Chem Soc 133:19358–19361

68. Pudney CR, Guerriero A, Baxter NJ, Johannissen LO, Waltho JP, Hay S, Scrutton NS (2013) Fast protein motions are coupled to enzyme H-transfer reactions. J Am Chem Soc 135(7): 2512–2517
69. Toney MD, Castro JN, Addington TA (2013) Heavy-enzyme kinetic isotope effects on proton transfer in alanine racemase. J Am Chem Soc 135(7):2509–2511
70. Świderek K, Javier Ruiz-Pernía J, Moliner V, Tuñón I (2014) Heavy enzymes: experimental and computational insights in enzyme dynamics. Curr Opin Chem Biol 21:11–18
71. Luk LYP, Loveridge EJ, Allemann RK (2014) Different dynamical effects in mesophilic and hyperthermophilic dihydrofolate reductases. J Am Chem Soc 136(19):6862
72. Luk LYP, Ruiz-Pernía JJ, Dawson WM, Loveridge EJ, Tuñón I, Moliner V, Allemann RK (2014) Protein isotope effects in dihydrofolate reductase from *Geobacillus stearothermophilus* show entropic–enthalpic compensatory effects on the rate constant. J Am Chem Soc 136(49):17317. doi:10.1021/ja5102536
73. Meinhold L, Clement D, Tehei M, Daniel R, Finney JL, Smith JC (2008) Protein dynamics and stability: The distribution of atomic fluctuations in thermophilic and mesophilic dihydrofolate reductase derived using elastic incoherent neutron scattering. Biochem J 94(12):4812–4818
74. Oyeyemi OA, Sours KM, Lee T, Resing KA, Ahn NG, Klinman JP (2010) Temperature dependence of protein motions in a thermophilic dihydrofolate reductase and its relationship to catalytic efficiency. Proc Natl Acad Sci U S A 107(22):10074–10079
75. Myllykallio H, Leduc D, Filee J, Liebl U (2003) Life without dihydrofolate reductase FolA. Trends Microbiol 11(5):220–223

Ab Initio Folding of Glycophorin A and Acetylcholine M2 Transmembrane Segments Via Simplified Environment Molecular Simulations

Eduardo V. Jardon, Peter J. Bond, and Martin B. Ulmschneider

Abstract The folding of glycophorin A and the acetylcholine receptor M2 domain from an extended conformation was studied by computer simulations. A classical molecular mechanics all-atom force field was combined with an implicit solute environment, which was modeled using the generalized Born theory of solvation. An efficient Monte Carlo sampling scheme that has demonstrated to yield results equivalent to molecular dynamics simulations was used in combination with a replica exchange algorithm to explore the folding of the peptides in aqueous and membrane environments.

In the membrane glycophorin A and M2 formed continuous membrane spanning helices for low temperature replicas. The lowest temperature replica folded the native trans-membrane helix with ~2 Å root mean square deviation compared to the experimental structures in detergent micelles. With increasing temperature the replicas exhibit a rapidly decreasing helicity, and the highest temperature replicas retain a completely extended conformation, in agreement with expected thermodynamic behavior.

After folding the helices show a large degree of conformational flexibility, while retaining their secondary structure. This indicates that the implicit system captures the essential components of the native membrane environment, while still allowing for adequate sampling of the accessible conformational space. Comparison with coarse-grain simulations yields good agreement.

In an aqueous environment the systems form conformers with a large degree of beta-structure, which is completely absent in the membrane system. These structures exhibit very large root mean square deviations from the native trans-membrane helix.

E.V. Jardon
Universidad Autónoma Metropolitana, Iztapalapa, Mexico

P.J. Bond
Bioinformatics Institute, 30 Biopolis Street, #07-01 Matrix, Singapore 138671, Singapore

M.B. Ulmschneider (✉)
Department of Materials Science and Engineering, Johns Hopkins University, Baltimore, MD 21218, USA
e-mail: martin@ulmschneider.com

L. Olivares-Quiroz et al. (eds.), *Physical Biology of Proteins and Peptides*,
DOI 10.1007/978-3-319-21687-4_7

Keywords Implicit membrane • Generalized Born • Protein folding • Monte Carlo • Coarse-grain • Acetylcholine • Glycophorin A

Introduction

Understanding membrane protein folding remains one of the most important and challenging problems of theoretical biophysics. Even though membrane proteins account for roughly one quarter of all proteins in most genomes [1–3] only ~400 atomic resolution structures are known to date [4].

Interactions with the lipid bilayer environment are crucial for correct membrane protein folding and function. The two-stage model for α-helical proteins postulates that helices first form across the membrane and assemble into a functional protein in a second step [5, 6]. Molecular simulations offer a potential route to elucidate this process.

Simulations of membrane proteins in their native membrane environment are usually limited to the sub-microsecond timescale due to the large number of nonbonded interactions that need to be evaluated for such complex systems. While this allows for the study of protein stability in a lipid bilayer or even self-assembly of protein/detergent micelles for various proteins [7, 8] it is usually too short to study protein folding (μs–ms timescale) or protein function (ms timescale).

Several approaches have been employed to reduce the computational effort and thus increase the timescales accessible to simulations. Coarse grain methods reduce the number of interactions by grouping atoms together into interaction sites comprising ~10 atoms [9]. The increase in sampling is achieved by the loss of atomic detail and the models need to be carefully parameterized to reproduce correct physical properties. While very useful for studying large scale systems [10] this method is not useful to study protein folding or function since these processes involve a complex interplay of intra and inter-molecular forces, which can only be captured correctly at the atomic scale. The second approach is to retain an all-atom description of the protein, while treating the environment implicitly. This has the advantage of combining large timescales with atomic detail, while largely avoiding parameterization issues by being able to build on the vast experience gathered with classical force fields. Its shortcomings are that the detail of protein-solvent interactions, particularly hydrogen bonds is lost. This can create simulation artefacts such as over-stabilized salt bridges.

Implicit solvation models generally treat the solvent as a dielectric continuum. For spherical ions in a homogeneous isotropic dielectric the solvation energy can be determined analytically as demonstrated by Born [11]. The generalized Born solvation model extends this equation to macromolecules, which are approximated as an assortment of charged spheres [12]. The immense success of this method in globular protein and peptide folding simulations (see e.g., [13–17]) has encouraged attempts to apply the generalized Born formalism to represent the membrane environment implicitly [18–25]. These models describe the membrane environment as a uniform hydrophobic slab and have been used successfully to fold and assemble small helical membrane peptides [22, 26–28].

In the present study we apply a computationally efficient implicit membrane model to study the folding of the transmembrane segments of glycophorin A (GpA) and the nicotinic acetylcholine receptor (M2). The model allows simulation of large systems and long timescales, while still being accurate enough to compare reasonably well with experimental data. Statistical analyses of membrane proteins [29] as well as translocon mediated insertion experiments of designed polypeptides [30, 31] suggest that the insertion energy of a residue at a certain position along the membrane normal is a property of its local solvation environment. It therefore seems reasonable to model a membrane as a smoothly varying solvation function along the bilayer normal. In the present study the membrane was treated as a region that becomes increasingly apolar (i.e., increasingly inaccessible to the solvent) towards the center of the membrane. However, rather than changing the dielectric constant as a function of the membrane normal the self-solvation energy of an atom, which accounts for the largest part of the solvation energy [32], was modified to vary smoothly between full solvation in bulk water and a limiting value for burial at the center of the membrane. At this stage of the development of the method any increased polarity at the charged bilayer interfaces was neglected. Nevertheless, the solvent exclusion properties of the membrane model account for the bulk properties of a lipid-bilayer. A general problem of membrane protein simulation lies in the correct treatment of charged residues at the membrane interfaces. In nature burial of a charged residue inside a membrane will almost certainly involve a change of protonation state or accompaniment by a hydration shell. Here the charged interfaces might play an important role, however, this is beyond the means of the present membrane model.

Another popular way of speeding up simulation of membrane proteins are coarse-grain models [9, 33, 34]. Very long timescales can be reached due to the reduced number of interaction sites. Both the implicit membrane and the coarse graining are attempts to overcome the sampling problems of classical molecular dynamics simulations. The two methods are complementary, allowing to study different aspects of the peptide-membrane system. Details of the folding process can be explored with in the implicit method, while the membrane is by construction fixed and imperturbable. The coarse-grain approach on the other hand restrains the secondary structure of the peptide in the folded state, but allows efficient simulation of lipid bilayer formation and perturbation. It is therefore interesting to study how both compare for the systems presented. To investigate this, we performed a series of coarse grain self-assembly simulations of M2 and monomeric GpA starting from a random box of lipid/water.

Methods

The Generalized Born Membrane

In 1920 Max Born showed that the solvation free energy of an ion is inversely related to the ionic radius [11]. This relationship can be utilized and generalized to represent the solvation free energy of a solute implicitly by treating a solute as a collection of ions with variable Born radii. For a given solute each atom will thus have a Born radius associated, which determines the contribution of that atom to the total solvation free energy of the solute. For example, an atom deeply buried inside a protein will have a very large Born radius, and hence a small contribution to the total solvation free energy of the protein, while an atom located on the solvent accessible surface of the protein will have a large contribution to the solvation free energy and hence a small Born radius. The method generally works by empirically estimating the Born radii from the degree of burial of a particular atom inside the solute [35]. The adaptation of this generalized Born (GB) method to include an implicit membrane has been described in detail in a previous publication [22]. The membrane was treated as a region in a uniform polar solvent with a dielectric constant $\varepsilon_{\text{water}} = 80$, that becomes increasingly inaccessible to the solvent towards its center. Both the protein interior and the membrane are assumed to have the same interior dielectric constant of $\varepsilon_{\text{membrane}} = 2$. The Born radii are calculated using the fast asymptotic pair-wise summation of Qiu and Still [35],

$$G'_{\text{pol},i} = \Gamma\left(z_i,, R_i,, L\right) + \sum_j^{1\,2} \frac{P_2 V_j\left(z_j\right)}{r_{ij}^4} + \underbrace{\sum_j^{1\,3} \frac{P_3 V_j\left(z_j\right)}{r_{ij}^4} + \sum_j^{1\geq 4} \frac{P_4 V_j\left(z_j\right) ccf}{r_{ij}^4}}_{\text{sums only involving atoms with } |z| > \text{L}}, \quad (1)$$

where P_1–P_4 are the parameters determined by Qiu et al. [35], and *ccf* is a close contact function. This method has been demonstrated to yield excellent results in predicting experimental free energies of solvation as well as hydration effects on conformational equilibria [36]. By modifying the pair-wise summation to solute atoms the self-solvation terms $\Gamma(z_i, \text{L})$ as well as the atomic volumes $V(z_i)$ were made to vary smoothly between full solvation and a limiting value for burial at the center of the membrane. We propose a Gaussian shape

$$\Gamma\left(z_i\right) = g_{\text{bulk}} + \left(g_{\text{center}} - g_{\text{bulk}}\right) e^{g\left(z_i^2 / L^2\right)}, \quad (2)$$

where g_{bulk} is the limiting value of Γ at a large distance from the membrane (i.e., z » L) corresponding to the self-solvation term of the unmodified generalized Born method $g_{\text{bulk}} = -166 / \left(R_i + offset + P_1\right)$, while g_{center} is the value of Γ at the membrane center. We used a Gaussian with $\gamma = -3.0$ and membrane half width of L = 15 Å, while $g_{\text{center}} = -7.67$ kcal/mol, as reported previously [18, 37].

The nonpolar part of the solvation free energy is modeled using an effective surface tension associated with the solvent accessible surface area (SA) [35]. Instead of a costly calculation of the accurate surface area, a mimic based on the Born radii is used, which has been shown to be very accurate, but much faster [38]. As it is moved towards the center of the membrane the surface energy contribution of each atom is scaled down by a Gaussian function of the same width as Γ. Thus for distances far from the membrane (i.e., z » L) the nonpolar contribution is included with the positive surface tension of solvation in water, while in the center of the membrane the surface tension is negative, (i.e., energy is gained by moving into this phase from the gas phase) as determined experimentally [39]. The surface tension contribution of each atom was varied using a Gaussian function with $\gamma = -1.5$, interpolating between the limiting values of 12 cal/mol Å^2 in bulk solvent and −19 cal/mol Å^2 at the membrane center.

It should be noted that the present membrane model neglects any effects due to differences in lipid composition and charge distribution of the two bilayer leaflets as well as effects due to the trans-membrane voltage. However, it is in principle possible to include these properties by replacing the Gaussians with an equivalent non-symmetric function.

No parameters were changed or optimized for the present simulations.

Monte Carlo Sampling

The simulations were run with a Monte Carlo program developed by the authors specifically for the simulation of proteins in a GB/SA continuum solvent. An efficient concerted rotation sampling technique [40] is used to move the protein backbone; in addition there are single rapid side-chain moves, with a ratio of three side-chain moves per backbone move. The potential energy was evaluated with the OPLS all-atom force field [41], and the Monte Carlo simulations used Metropolis sampling. All non-bonded interactions as well as the generalized Born energy were truncated using a cutoff of 14 Å, and the Born radii were recomputed for every configuration. The described setup has been shown to perform well in sampling DNA [42] and protein folding simulations [17], where the native state of several polypeptides was rapidly determined starting from extended conformations. We recently demonstrated that this method is equivalent to molecular dynamics sampling, with both methods able to find the native state of several polypeptides with comparable computational effort [43].

Simulation Setup

The membrane spanning segments were modeled in completely extended conformation and simulated in an implicit membrane as well as aqueous solvent. The termini were acetylated and amidated respectively. In the implicit membrane system the peptide was arranged so that it spans the membrane (see Fig. 1). The present implicit membrane generalized Born algorithm requires the protein interior dielectric constant to be equal to the membrane dielectric constant. In accordance with previous work a value of $\varepsilon_{protein} = \varepsilon_{membrane} = 2$ was chosen for the solvation energy calculations of the membrane systems (TM). To investigate whether the change of the protein interior dielectric constant has an effect on protein conformations the water simulations were performed with $\varepsilon_{protein} = 1$ (W1) as well as $\varepsilon_{protein} = 2$ (W2). The water dielectric constant was kept at $\varepsilon_{water} = 80$ for all systems. The modeling and simulation software is available freely at www.biowerkzeug.com.

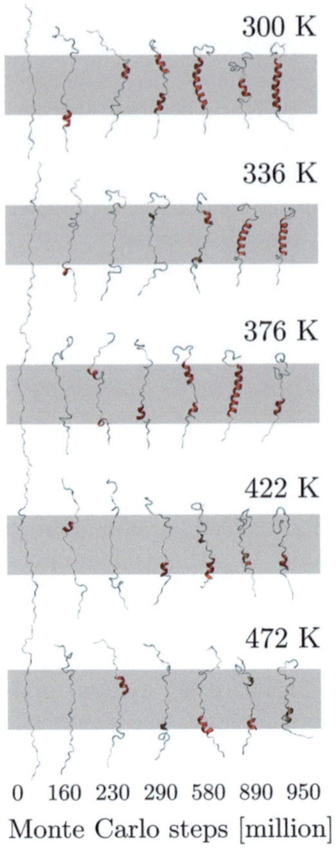

Fig. 1 Trans-membrane folding of the glycophorin A monomer as a function of temperature. Only every second replica is shown for clarity. The lowest four temperature replicas (300, 318, 336, and 356 K) fold into stable membrane spanning helices within the first ~400 million Monte Carlo steps, while the higher temperature replicas retain a largely extended conformation. It should be noted that the implicit membrane does not represent a hydrophobic slab but rather a Gaussian shaped hydrophobic zone, thus the 30 Å slabs shown are for reference only

Parallel Tempering

The replica exchange method has recently been reviewed in detail [44]. Ten replicas of each system were set up with identical fully extended initial configuration and exponentially spaced temperatures in the range 300–500 K, namely: 300, 318, 336, 356, 376, 398, 422, 446, 472, and 500 K. Every 10^4 Monte Carlo moves a replica swap with transition probability

$$p_{1\to 2} = \exp(-\Delta), \tag{3}$$

where

$$\Delta = \left(\frac{1}{kT_1} - \frac{1}{kT_2}\right)(E_1 - E_2) \tag{4}$$

is attempted. E_1 and E_2 are the total energies of two conformers at temperatures T_1 and T_2 respectively. High temperature replicas facilitate the crossing of energy barriers, while low temperature replicas extensively sample low energy conformations.

This enables the efficient and increased sampling of the entire system by frequent crossing of high energy barriers.

The paper refers to two types of trajectories. The first is the constant temperature replica, e.g., T=300 K, in this trajectory the structure is frequently swapped while the simulation temperature is always exactly 300 K. The second trajectory follows the progress of an individual starting structure, e.g., the initial conformation of the 300 K replica as it changes temperature, being melted or cooled.

Analysis

A cluster analysis, using the pairwise method of Daura et al. [45], was carried out to assess the main secondary structural motifs populated during the simulations. Since pairwise clustering becomes computationally costly for large coordinate sets, the structures were taken every millionth MC step and were superimposed using main-chain least-square superimposition with a similarity cutoff of 2.5 Å. Since most clusters are sparsely populated and do not represent the main features of the simulations, only the most populated clusters were further analyzed.

The free-energy was calculated as a function of the root-mean-square Cα deviation (RMSD) from the experimental structure (PDB code: 1afo, model 1). For a system in thermodynamic equilibrium, the change in free energy on going from one state to another is given by

$$\Delta G = -RT \ln \frac{p_1}{p_2} \quad (5)$$

where R is the ideal gas constant, T is the temperature and p_i is the probability of finding the system in state i. The RMSD was divided into a grid with a spacing of 0.3 Å, and the free-energy was calculated for each bin. In addition, the average total system energy E was calculated for each cell, which is the sum of peptide internal energy plus the interactions with the solvent and membrane given by the GB/SA solvation free energy. In order to check the effect of a different grid size on the thermodynamic properties, different grid spacings of 0.1– 0.5 Å were used to construct the same free-energy landscapes. Surfaces constructed on smaller bin sizes tended to be rougher as fewer points are available per bin, but all are similar, with the same overall shape and spread of the free-energy.

Results and Discussion

Trans-membrane and Aqueous Folding of Glycophorin A

Glycophorin A (GpA) is a 131 residue protein that resides in the human red blood cell membrane. It consists of a single trans-membrane helix, which forms stable homodimers in detergent micelles (PDB code: 1afo) [46] and lipid bilayers [47].

GpA self-association has been studied extensively by experimental and computational tools since it represents a model two-stage folding protein (see [48] and [49] for recent reviews). The membrane spanning segment of glycophorin A (residues Val62–Lys101) forms native state trans-membrane helical conformations in the absence of the remainder of the peptide. In the present work we concentrate on the first step of the two-stage pathway, the folding of a trans-membrane helical monomer.

Figure 1 shows the folding progress of the trans-membrane system over the 1×10^9 MC step simulation. Only five of the ten replicas are displayed for clarity. The lowest four temperature replicas (300–356 K) fold into continuous trans-membrane helices. Replicas with a temperature higher than 356 K (i.e., 376–500 K) retain a large amount of extended conformations and do not form stable membrane spanning helices. Nevertheless, they contain a substantial amount of helical secondary structure. Particularly at the interfacial regions partial helices form frequently, while the center of the membrane spanning segment is generally unfolded. There is virtually no beta structure except occasionally at the interfaces. Interestingly the helix formation of the lower replicas initiates at the interfaces as well. The ability of the interfacial regions to provide a strong catalyst for secondary structure formation is well known and generally attributed to its amphiphilic nature [50, 51].

All folding events take part in the first 500 million MC steps. To quantify the similarity to the native state, we aligned each conformation to the Cα positions of the helical segment (Glu72–Ile99) of the experimental structure and calculated the RMSD. Figure 2 shows that the RMSD with respect to the experimental structure (1afo, model 1) decreases sharply in that period and then levels off in the second half.

In water an initial structural collapse lowers the RMSD in the first half of the simulation. Later the RMSD increases again, as the peptide moves from a compact collapsed state into an ensemble of partially folded conformations. This effect is more pronounced for the simulation with $\varepsilon_{\text{protein}} = 2$. Figure 3 plots the average helicity for each residue as a function of replica temperature for the last 500 million MC steps. As opposed to the membrane simulation the water simulation displays little helical secondary structure (Fig. 3, panel W1). Increasing the protein interior dielectric constant from $\varepsilon_{\text{protein}} = 1$ to $\varepsilon_{\text{protein}} = 2$ increases the propensity of the peptide to fold helical secondary structure in water (Fig. 3, panel W2), by lowering the penalty for hydrogen bond burial (see "Methods"). Nevertheless, the pattern of helical residues is different from the membrane simulation (Fig. 3, panel TM).

The membrane systems exhibit strongly increasing helicity at lower temperatures. This is particularly pronounced in the core region, while interfacial regions retain a large degree of helicity at elevated temperatures. In contrast, both water systems show little to no temperature dependence in their propensity to form helices.

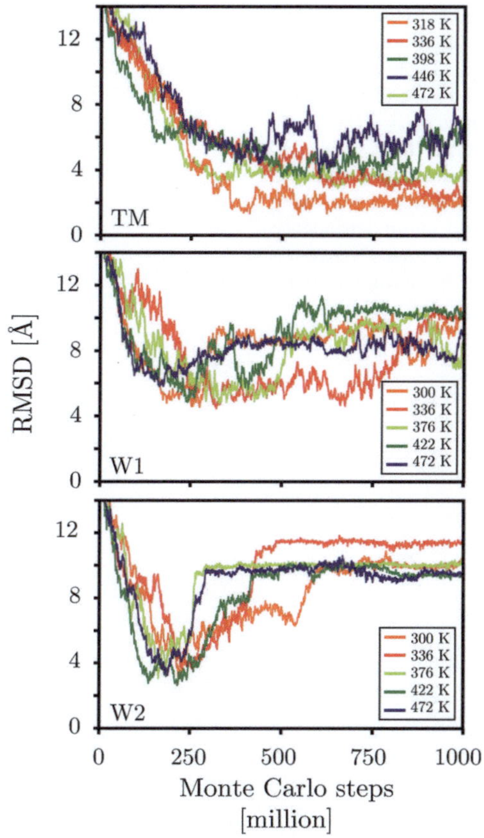

Fig. 2 Root mean square deviation from the glycophorin A NMR structure (1afo, model 1) plotted against simulation time for various initial starting structures. The equivalent graph for constant temperature trajectories is similar but displays large discontinuous jumps in the RMSD due to structures being swapped, hence individual structures were followed for clarity. *TM* membrane simulation, *W1* water simulation with protein interior dielectric constant $\varepsilon_{protein} = 1$ and *W2* water simulation with $\varepsilon_{protein} = 2$

Cluster Analysis

All runs show an immediate relaxation of the system from the extended state. This initial fast collapse takes around 300 million MC steps. In MC simulations straight downhill folding from a high-energy extended conformation to a compact equilibrium-like structure, enhanced in this case by the use of implicit solvent, is achieved by converging on the minimum only after many accepted and rejected moves. This gives the trajectories their rattle-like appearance. Once collapsed conformations are reached the trajectories show frequent transitions between compact folded states. Multiple folding/unfolding cycles are observed and the system can be trapped in compact coil or β-hairpin conformations for considerable time, usually with several backbone hydrogen bonds formed. A cluster analysis of the second half ($0.5–1 \times 10^9$ MC steps) of the lowest temperature trajectory (T = 300 K) reveals the main secondary structure motifs sampled after this initial collapse (see Table 1). Most of the simulation time is spent in a membrane spanning helical conformation, as observed in the NMR

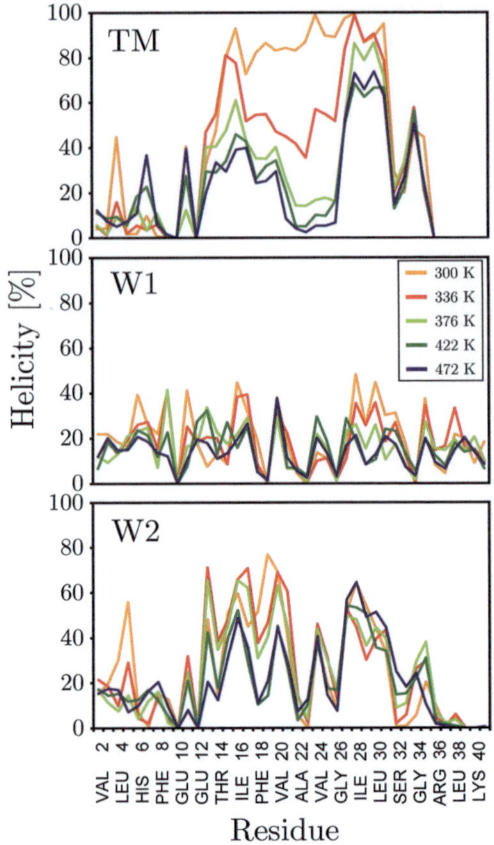

Fig. 3 Average helicity versus residue for various replica temperatures of the membrane (TM), water with $\varepsilon_{\text{protein}} = 1$ (W1) and water with $\varepsilon_{\text{protein}} = 2$ (W2) systems. The membrane system displays a strong temperature dependence. The averages were over the last 500 million Monte Carlo steps

experiments [46]. The number of turns varies between 4 and 7 and the helix displays a large degree of conformational flexibility with straight (cluster 2), curved (clusters 1 and 5) and kinked (cluster 5) conformations. Sometimes an unfolded structure is passed down from higher temperature replicas (cluster 4). Over 50 % of conformations are part of a large number (79) of less populated clusters (occupancy<4 %), indicating that the conformational space has sampled well.

Tables 2 and 3 show the equivalent cluster analyses of the water simulations. The most striking difference is the large degree of beta-structure, which is virtually absent in the membrane systems. Although the amount of helical secondary structure increases for the W2 simulation the native state is not found by any of the replicas.

Free Energy Surfaces

To obtain insights into the folding mechanism, the many-dimensional system was projected onto a structural coordinate. Figure 4 shows the free-energy and total energy of the lowest temperature replica of each simulation as a function of

Table 1 Cluster analysis of the lowest temperature (T = 300 K) replica for the second half (500–1000 MC steps) of the membrane simulation

Cluster	Structure	Occupancy (%)
1		17.0%
2	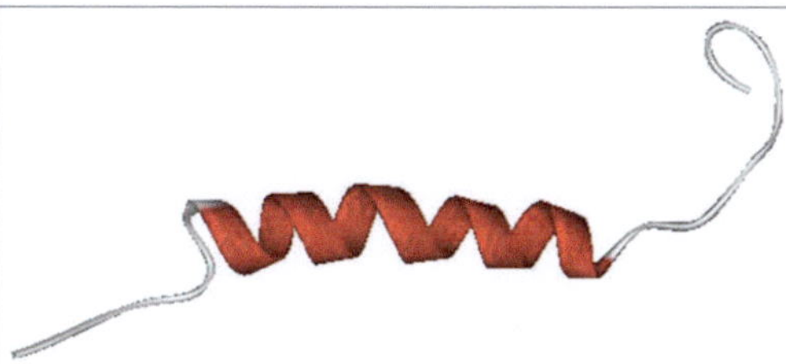	15.8%
3		5.6%
4	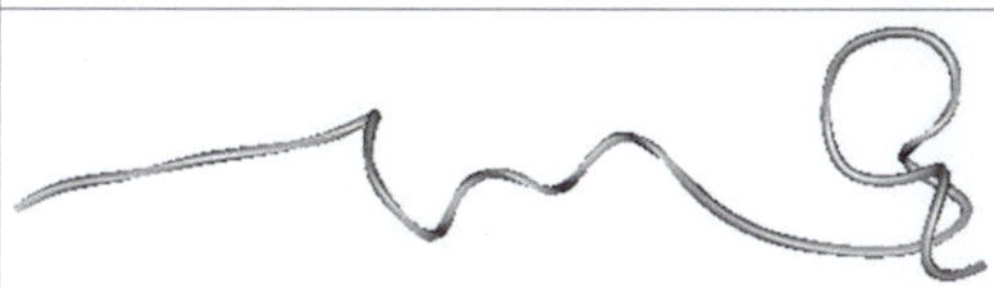	5.1%
5	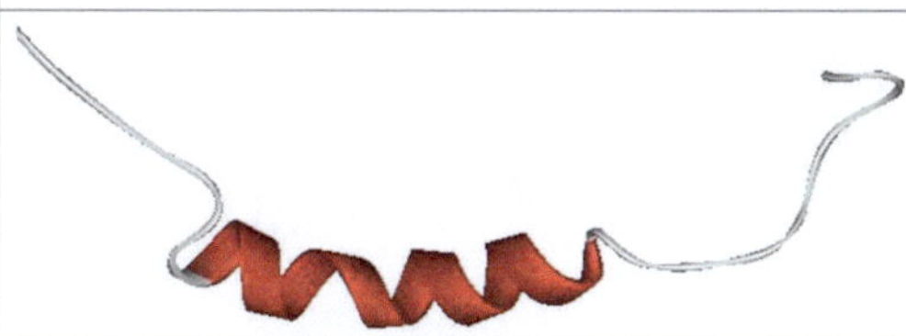	4.6%
6–85	Coils, helices, partially folded helices	51.8%

The protein interior as well as the membrane have a dielectric constant of ε = 2. The cluster RMSD cut-off was 2.5 Å

backbone RMSD from the NMR structure (PDB code: 1afo, model 1). A two-state folding pattern is evident, with a deep valley for the folded states at ~1.5–3.5 Å, corresponding to a continuous membrane spanning helix, and a shallow broad basin of the unfolded structures ranging from 8 to 11 Å. The trans-membrane helix also exhibits a low total energy. The magnitude of the folding barrier in this landscape is ~3 kcal/mol, and the native state is stabilized with ΔG ~ 1 kcal/mol compared to the unfolded basin.

The aqueous simulations exhibit a broad basin of structurally unrelated unfolded states (cf. Tables 2 and 3). Figure 4 shows that the basins range from 5 to 10 Å for the W1 and from 8 to 10 Å for the W2 system. In the case of W1, which only samples to within 5 Å of the native state this unfolded basin also contains the lowest

Table 2 Cluster analysis of the lowest temperature (T = 300 K) replica for the second half (500–1000 MC steps) of the water simulation

Cluster	Structure	Occupancy (%)
1	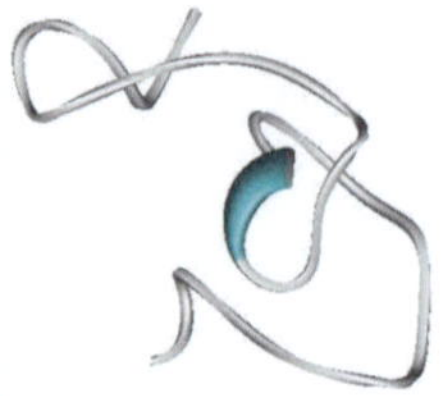	17.5%
2	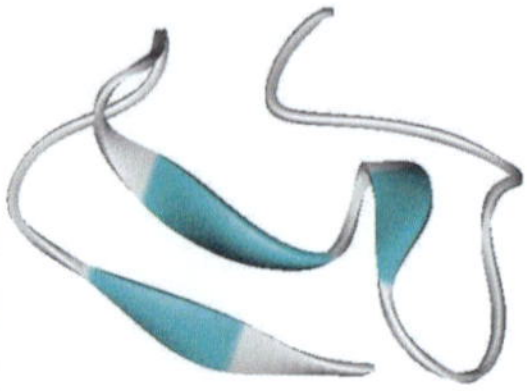	12.0%
3	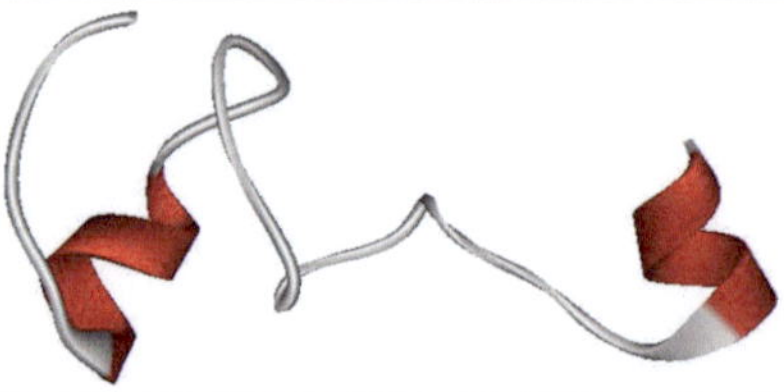	7.4%
4	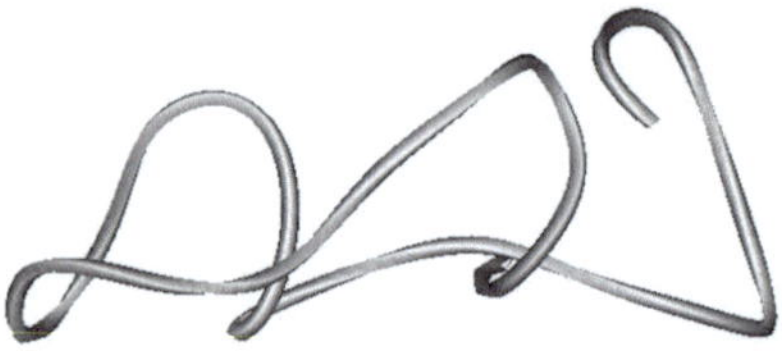	7.0%
5	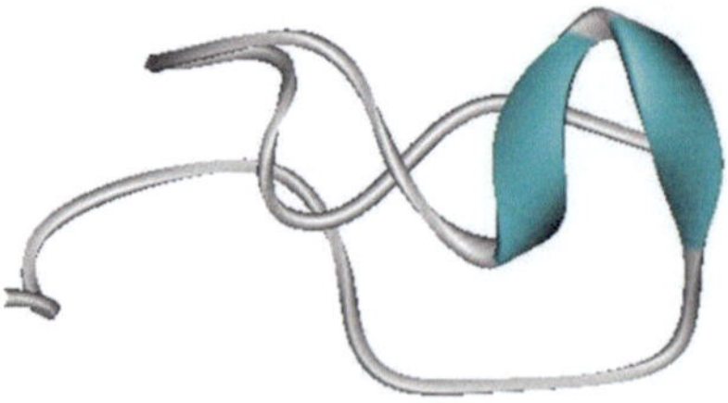	4.3%
6–108	Coils, partial helices, hairpins	51.8%

The protein interior dielectric constant was $\varepsilon = 1$

energy structures. Interestingly the W2 simulation samples the conformational space to with 2.5 Å, and this structure also has the lowest energy, but carries a ~2 kcal/mol free-energy penalty compared to the unfolded basin. In general the increase in protein interior dielectric constant results in a slight increase in the formation of alpha as well as beta secondary structure.

Table 3 Cluster analysis of the lowest temperature (T = 300 K) replica for the second half (500–1000 MC steps) of the water simulation

Cluster	Structure	Occupancy (%)
1	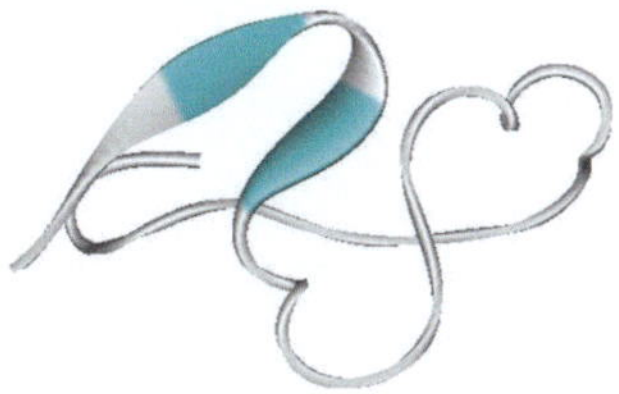	27.6%
2	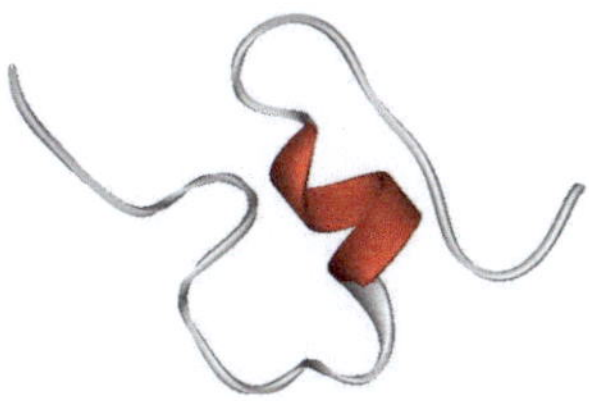	25.3%
3	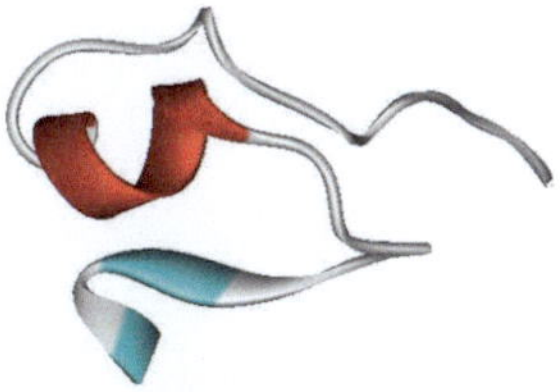	9.1%
4	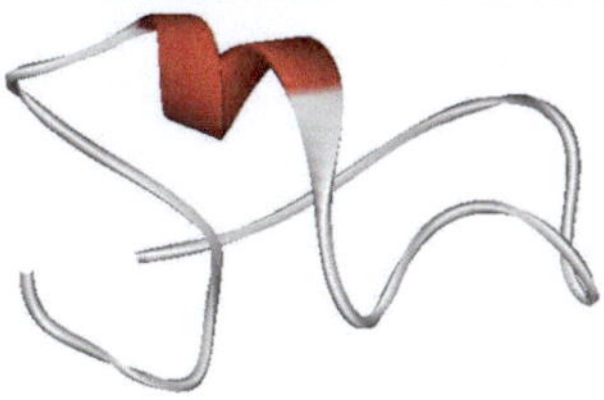	8.9%
5	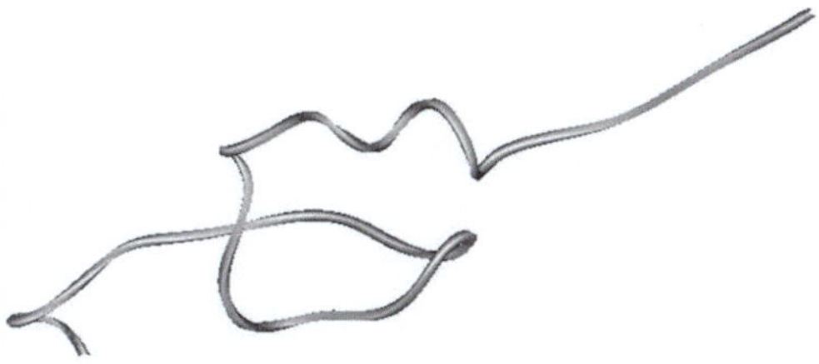	7.7%
6–44	Coils, partial helices, hairpins	16.9%

The protein interior dielectric constant was ε = 2

Structural Stability of the Trans-membrane Configuration

Figure 5 shows the helicity, tilt angle and center of mass movement along the membrane normal of the peptide as a function of simulation time. Low temperature replicas exhibit steady folding of the trans-membrane helices within the first 500

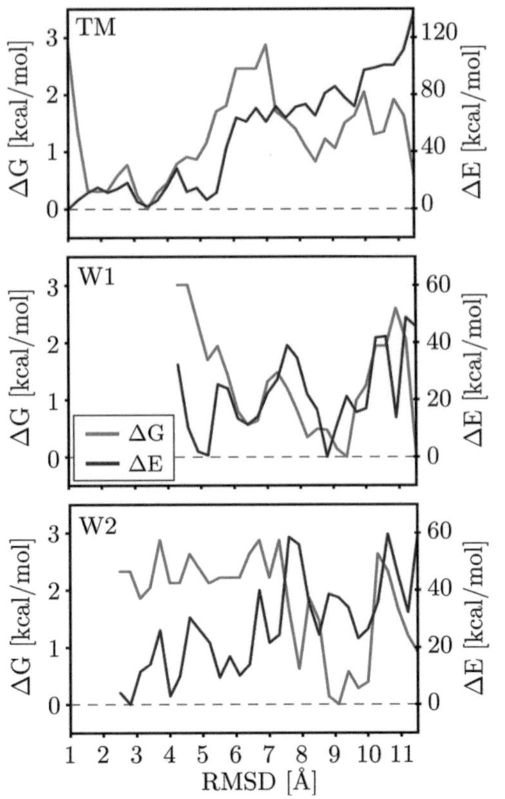

Fig. 4 One dimensional free-energy (ΔG) and force field energy (total energy ΔE) curves for the lowest temperature replica (T = 300 K) of the membrane (TM), water with $\varepsilon_{protein} = 1$ (W1) and water with $\varepsilon_{protein} = 2$ (W2) simulations. The lowest states encountered were defined as the zero potentials

million MC steps. Higher temperature replicas stay further away from the native state and do not form stable TM helices. This result is contrary to a previous simulation with a similar implicit membrane setup that reported all 16 replicas ranging from 300 to 600 K folding into continuous helices [26]. The authors suggest this is most likely due to a strong helical bias of the CHARMM force field [52].

The helices show a tendency to tilt frequently in the range 0–30° regardless of temperature, demonstrating that the system retains a large conformational flexibility even at low temperatures. The mean tilt angle of the 300 K replica of 17 ± 7° is slightly larger than the 8–10° ± 3–4° monomer tilt angles reported from 1.5 ns all-atom stimulations of a folded GpA monomer in various explicit bilayers with different lipids [53].

GpA displays considerable movement along the membrane normal, with the center of mass of the helices oscillating between 5 and –10 Å. However, due to the strongly charged termini all systems remain firmly integrated into the membrane. Nevertheless, the movement is much more pronounced than the ±2 Å found for folded GpA in explicit bilayers [53].

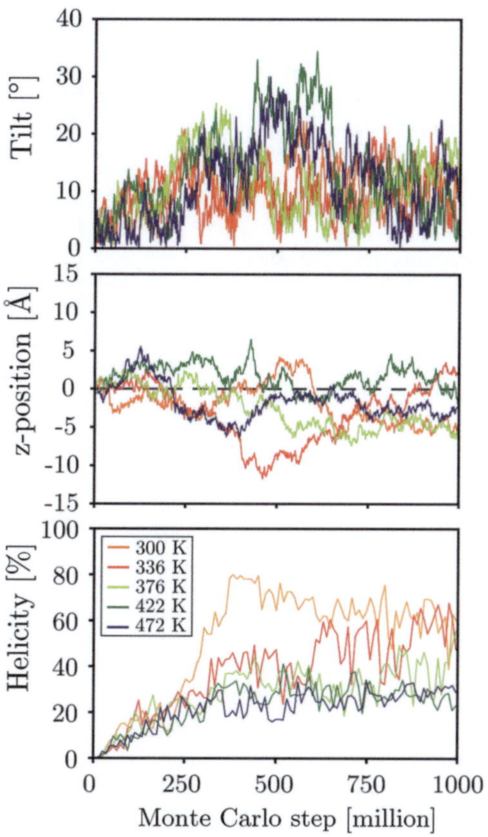

Fig. 5 Tilt angle, center of mass z-position and helicity versus simulation time for various temperature replicas of the membrane systems

Sampling Efficiency and Estimates of Folding Times

Figure 6 shows that the temperature swapping rate of replica exchange simulations equilibrate quickly and level out at four accepted swaps per 10^6 MC moves, irrespective of system temperature. This indicates that the replicas are spaced with roughly equal phase-space overlap. The figure graphically demonstrates that each replica samples the whole temperature space of the simulations (only the 300, 398, and 500 K are shown for clarity, the others are exactly similar). Nevertheless, as a particular starting structure folds into a low free-energy conformation (e.g., a trans-membrane helical conformation) it remains in the low temperature range of the replicas. Contrary a structure that retains an unfolded or extended conformation is 'pushed up' towards the high temperature replicas, where it remains until if forms low energy secondary structure, which in turn will initiate its descend on the temperature ladder. This means that replicas that are 'stuck' in a non-native conformation are slowly pushed up the temperature ladder and consequently melted since structures closer to the native state are being passed down.

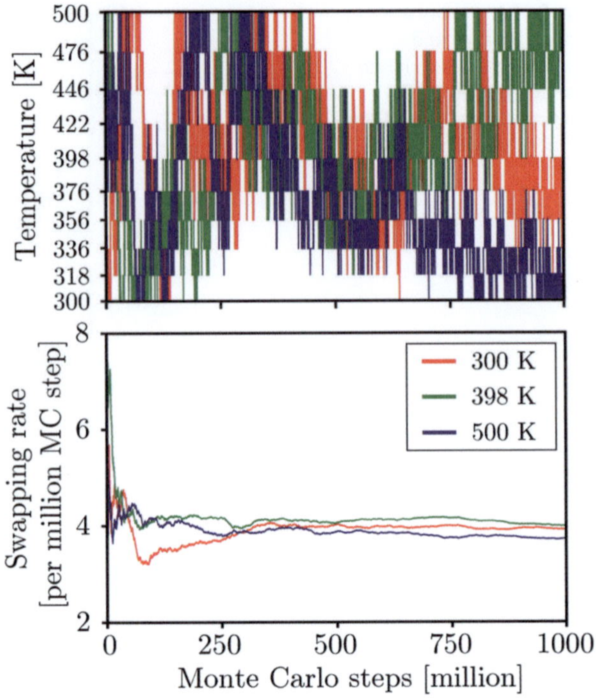

Fig. 6 *Upper panel*: Temperature versus simulation time for the 300, 398 and 500 K starting simulations of the membrane system. *Lower panel*: Swapping rate for the above starting structures versus simulation time. The other replicas behave identically and have been omitted for clarity. The respective data for the water systems is exactly similar

Folding times are very difficult to estimate correctly due to the low number of folding events, which leads to errors that are quite large. For example, in a previous, much shorter MC study of GpA using an implicit membrane with a surface tension solvent approximation, folding times in the range of 10^3 MC steps were encountered [54]. Results from a previous study in a generalized Born implicit membrane with a similar setup, but different force field and sampling method found folding in the ~100 ps timescale, roughly equal to 2×10^6 MC steps [26], significantly lower than the ~3.6×10^8 MC steps reported here, which correspond to a folding time of around 18 ns. All of these seem too short compared to experimental estimates from globular helices, which are in the 0.2 μs–ms range [55, 56].

The problem is that results are dominated by fast '*lucky collapses*'. This reveals a problem with starting simulations from extended conformations (in order not to bias results) rather than the equilibrated compact ensemble of the denatured state. In these simulations, the folding from high-energy extended conformations proceeds without crossing any barrier straight to the folded state, which is a special case that is unlikely to occur when initiating folding from low-energy compact structures. It has been noted before that fast folding events are not representative of the major folding pathways [57, 58]. Such lucky collapses are also responsible for the prominent well in the RMSD graphs of the W2 simulations (see Fig. 2), which come within 2.4 Å of the native trans-membrane configuration before unfolding into their preferred aqueous ensembles, indicating that simulations need to be sufficiently lengthy to be sure the conformations are converged.

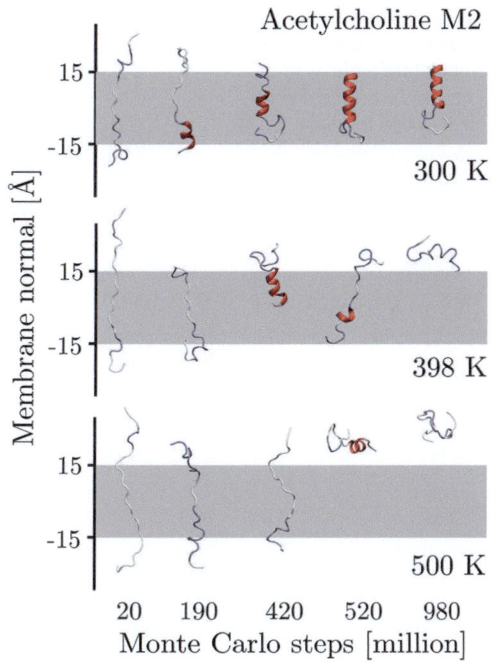

Fig. 7 Trans-membrane folding of the acetylcholine M2 domain monomer as a function of temperature. Only three replicas are shown for clarity. The lowest four temperature replicas: 300, 318 (*not shown*), 336 (*not shown*), and 356 K (*not shown*) fold into stable membrane spanning helices within the first ~400 million Monte Carlo steps, while the higher temperature replicas retain a largely extended conformation. It should be noted that the implicit membrane does not represent a hydrophobic slab but rather a Gaussian shaped hydrophobic zone, thus the 30 Å slabs shown are for reference only

The problem is aggravated by the use of replica exchange methods, since the fastest folding event in the many replicas is reported as the folding time. Therefore the more replicas are used the less reliable the folding time becomes. Thus our folding time of ~18 ns is most likely an underestimate of the real folding time.

Acetylcholine Receptor M2

The nicotinic acetylcholine receptor (AChR) is a neurotransmitter-gated ion channel responsible for the transmission of signals in the postsynaptic membrane of neurons. Of particular interest is the highly conserved M2 domain of the protein, an amphipathic α-helix that lines the interior of the aqueous pore. The M2 monomer has been studied in detail by Opella et al. using solution and solid-state NMR in dodecylphosphocholine (DPC) micelles and dimyristoyl-phosphatidylcholine (DMPC) bilayers, revealing a TM bound α-helix tilting ~12° with respect to the bilayer normal [59]. There is a large number of previous simulation work available on the M2 system, including MD studies [60, 61] and numerous MC simulations that combine an implicit membrane model with a reduced protein representation [62–64].

REMC simulations were run in a similar fashion to the glycophorin A runs with ten replicas for 1×10^9 MC steps each, starting from completely extended conformations perpendicular to the membrane plane. In addition, an identical REMC run was performed in pure water. Figure 7 shows the folding progress over the course of the simulation. Quick formation of α-helical conformations in TM orientation is

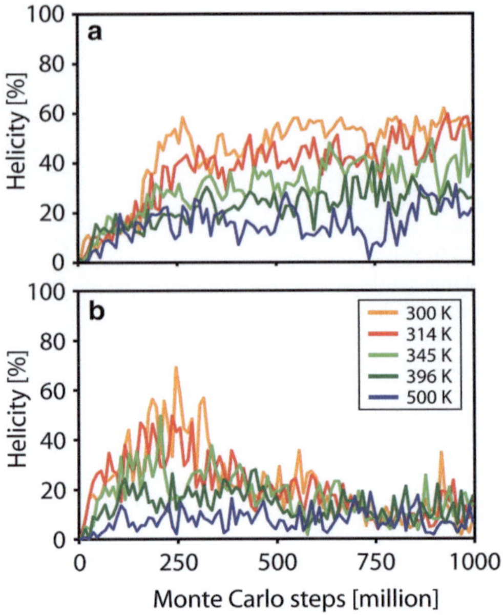

Fig. 8 Helicity versus simulation time for various temperature replicas of the acetylcholine M2 system. Panel (**a**): in implicit membrane. Panel (**b**): in water with $\varepsilon_{protein} = 1$

observed, while the highest temperature replicas sample mostly unfolded structures. The growth of helicity is visualized in Fig. 8 (upper panel) for a series of replicas. Similar to the NMR experiments, M2 is found to form a TM helix for all residues. Figure 9 shows the average helicity for each residue for several replicas, averaged over the last 700×10^6 MC steps. A strong dependence of helicity on temperature is evident, with only the 300 and 314 K replica having fully formed helices. Frequent replica swaps with partly folded structures result in a lower average helicity for Leu19–Arg25 for the 300 K replica, indicating the averages are not completely converged even after 1×10^9 MC steps. The corresponding graph for the pure water run is shown in the lower panel of Fig. 9. There is no significant helicity for any of the residues at all temperatures, suggesting that helical conformations are not stable outside of the membrane. Figure 8 (lower panel) shows the total average helicity as a function of simulation time for the pure water run. Some helical conformations are sampled during the initial collapse from the extended starting structure, but quickly dissolve for the rest of the simulations.

Once formed, the TM helix remains firmly inserted into the membrane. Figure 10 shows the tilt angle and center of mass z-position over the course of the 300 K replica. Fully helical structures are frequently swapped with only partly helical conformations resulting in considerable flexibility of the z-position as well as the tilt of the system. The strong spikes correspond to a swap into a completely unfolded conformation that resides on the membrane surface. While the REMC algorithm permits such transitions, the simulations immediately return to the energetically much more favored native structure in the next swap move. Thus, the contribution of the unfolded conformations to the thermodynamic averages is almost zero. The average z-position is 3.4 ± 3.9 Å and the helical domain shows a tendency to tilt

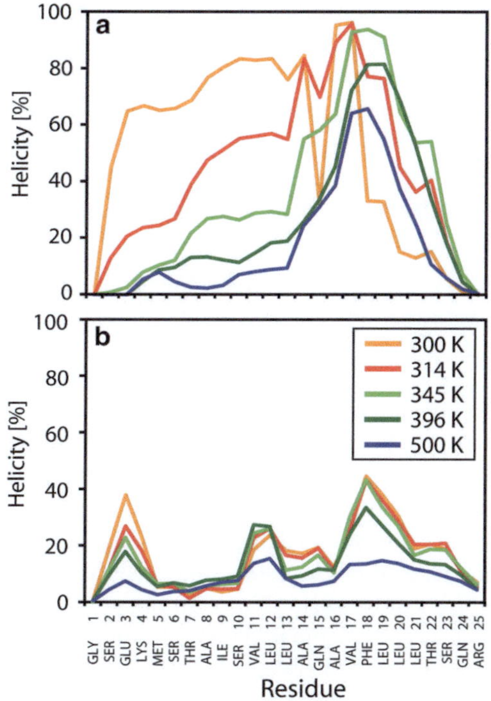

Fig. 9 Average helicity versus residue for various replica temperatures of the membrane (panel **a**) and water with $\varepsilon_{\text{protein}} = 1$ (panel **b**) systems. The membrane system displays a strong temperature dependence, while the water runs show only residual helicity at all temperatures. The averages were over the last 700 million Monte Carlo steps

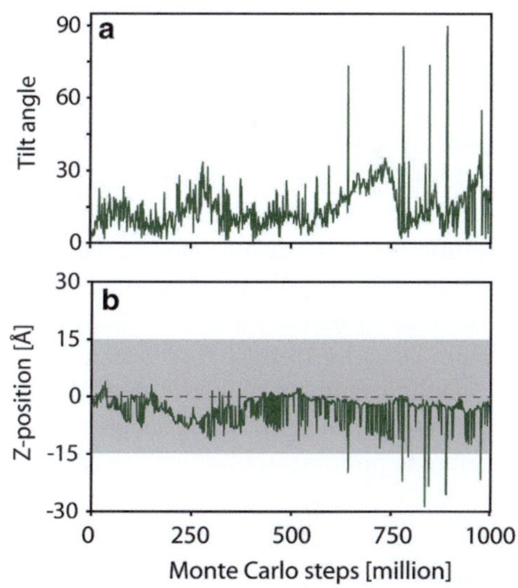

Fig. 10 Tilt angle and center of mass z-position versus simulation time for the 300 K replica of the acetylcholine M2 system. Replica swaps are visible as strong transitions. The largest spikes correspond to a swap into a replica in which the system is randomly coiled and rests on the membrane surface. There is only a very tiny contribution of such conformations to the 300 K averages, as the fully formed TM helical structures dominate

frequently in the range 0–30°, with a mean tilt angle of 15 ± 8°. This is very close to the ~12° reported in the experiments [59]. Thus the overall agreement with experiment is excellent, especially given the fact that no part of the implicit membrane model was tuned for any of the work presented here. Interestingly, the tilt angle matches exactly the 14–15° that were found in previous computational studies of the acetylcholine M2 system although a completely different model was used there [64, 65]. This model combined empirical free energies associated with lipid and peptide degrees of freedom with an electrostatic implicit membrane component that was calculated using a finite difference Poisson-Boltzmann solver.

Comparison with a Coarse Grain Model

The hydrophobic central sequences of the peptides were restrained to their folded conformations. Several simulations were run for 200 ns at 323 K, to be above the DPPC phase transition temperature. One simulation of each peptide resulted in the helix being stably inserted into a bilayer, and these were further analyzed. It should be noted that no lipid phase transitions are present in the implicit membrane model, which is by design always in the fluid phase.

Figure 11 shows snapshots of the membrane inserted peptides, after bilayer formation and insertion, which in both cases took a few tens of nanoseconds. While there is little to no deformation of the bilayer around the peptide is observed for M2, the termini of GpA sometimes rearrange to cause local deformation of the bilayer. Once stably inserted into the bilayer the overall motion with respect to the membrane can be compared for both simulation methods. In the implicit membrane, both peptides tilted between 0° and 30°, with means at 300 K of $17 \pm 7°$ and $15 \pm 8°$ for GpA and M2 respectively. In the coarse grain model, at 323 K, we also found tilting between ~0° and 30° and comparable means of $19 \pm 8°$ and $14 \pm 7°$ for GpA and M2 respectively (see Fig. 12). The agreement of the two membrane models is remarkable, given their substantial differences. Once inserted the helices remained firmly integrated into the lipid bilayer. As in the implicit method the centre of mass of the peptides oscillates substantially along the membrane normal, with GpA having $z = -2.9 \pm 2.2$ Å, and M2 $z = 2.6 \pm 2.0$ Å (see Fig. 12). This corresponds nicely to the implicit membrane results ($z_{GpA} = -2.0 \pm 2.5$ Å and $z_{M2} = 3.4 \pm 3.9$ Å) and confirms an insertion bias towards one side of the peptide.

Ultimately, the all-atom accuracy of the implicit membrane model is required for meaningful structure prediction that can be compared to experiments. But the equilibrium behavior of fully folded motifs seems to be equally well described by both the implicit membrane model and the coarse grain model. These results are very encouraging and open the possibility of studying much longer timescale events in membrane protein simulations than possible with a fully detailed descriptions.

Conclusion

The present work demonstrates that an implicit generalized Born membrane in combination with all-atom force fields and an efficient Monte Carlo sampling scheme can reliably predict the native state of membrane spanning peptides. By using a replica exchange algorithm the number of barrier crossing events was increased to yield reasonably converged configurational averages at much shorter timescales than

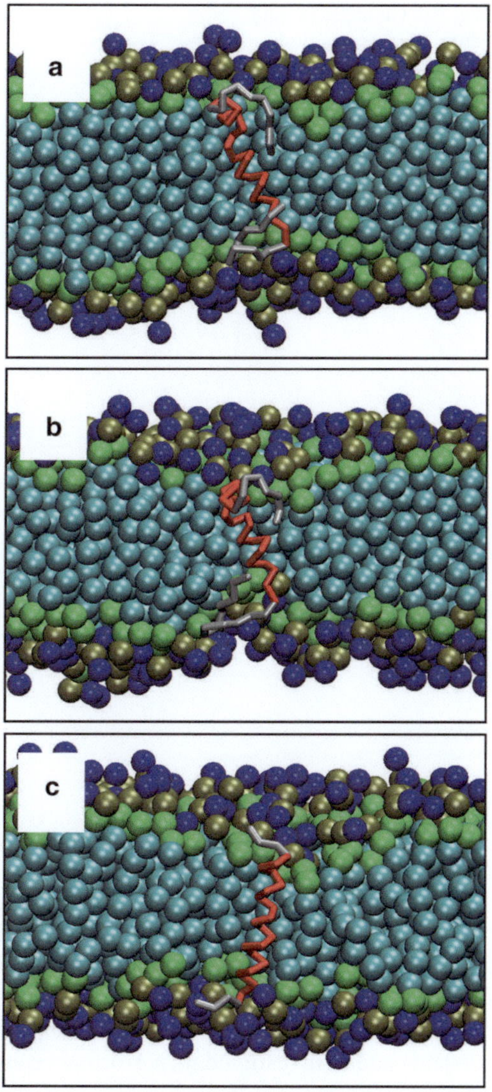

Fig. 11 (**a**) Snapshot of the coarse-grain simulation of the glycophorin A monomer with a typical conformation. (**b**) Conformation induced bilayer thinning around glycophorin A. (**c**) Coarse-grain simulation of acetylcholine M2

would be required otherwise. Nevertheless, the total simulation effort involved five simulations with ten replicas each giving a accumulated simulation time of 5×10^{10} MC steps, corresponding to ~2.5 μs of molecular dynamics [43].

The membrane simulations folded continuous membrane spanning helices with the low temperature replica sampling conformations to within 1.3 Å of the native state. In contrast the aqueous simulations formed stable beta-structures, that are completely absent in the membrane. This demonstrates that the membrane environment is crucial to determining the correct fold of the protein. Since the membrane is in essence modeled as an aqueous solvent excluded region it is reasonable to assume that desolvation in the hydrophobic membrane core is indeed the major property driving membrane protein folding.

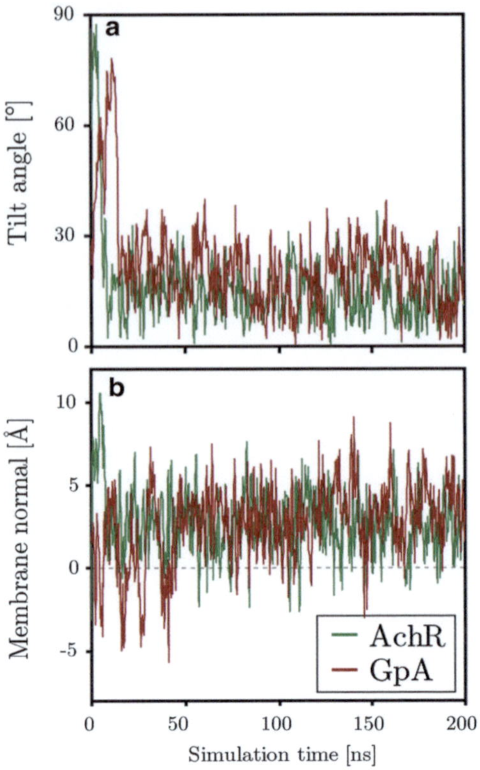

Fig. 12 Tilt angle (**a**) and center of mass position along the membrane normal (**b**) versus simulation time for coarse-grain simulations of acetylcholine M2 (AchR) and glycophorin A (GpA). The membrane centre is indicated

Once the helices have formed in the low temperature replicas they remain stably integrated into the membrane. Nevertheless, they retain a large conformational flexibility exploring the vicinity of the native state by frequent tilting, kinking and bending.

Increasing the protein interior dielectric constant from 1 to 2 had only a very limited effect on the aqueous simulations, slightly increasing the propensity of the protein to form secondary structure.

It will be of interest to apply the implicit generalized Born method to larger systems and to investigate its performance and accuracy compared to all-atom lipid bilayer membranes as well as coarse grain approaches.

Acknowledgement This research was funded by an International Fellowship of the Wellcome Trust (MBU).

References

1. Wallin E, Von Heijne G (1998) Genome-wide analysis of integral membrane proteins from eubacterial, archaean, and eukaryotic organisms. Protein Sci 7:1029–1038
2. Jones DT (1998) Do transmembrane protein superfolds exist? FEBS Lett 423:281–285

3. Krogh A, Larsson B, Von Heijne G, Sonnhammer EL (2001) Predicting transmembrane protein topology with a hidden Markov model: application to complete genomes. J Mol Biol 305:567–580
4. White SH (2004) The progress of membrane protein structure determination. Protein Sci 13:1948–1949
5. Popot JL, Engelman DM (1990) Membrane-protein folding and oligomerization—the 2-stage model. Biochemistry 29:4031–4037
6. Engelman DM, Chen Y, Chin CN, Curran AR, Dixon AM, Dupuy AD, Lee AS, Lehnert U, Matthews EE, Reshetnyak YK, Senes A, Popot JL (2003) Membrane protein folding: beyond the two stage model. FEBS Lett 555:122–125
7. Braun R, Engelman DM, Schulten K (2004) Molecular dynamics simulations of micelle formation around dimeric glycophorin A transmembrane helices. Biophys J 87:754–763
8. Bond PJ, Cuthbertson JM, Deol SS, Sansom MS (2004) MD simulations of spontaneous membrane protein/detergent micelle formation. J Am Chem Soc 126:15948–15949
9. Marrink SJ, Devries AH, Mark AE (2004) Coarse grained model for semiquantitative lipid simulations. J Phys Chem B 108:750–760
10. Bond PJ, Sansom MS (2006) Insertion and assembly of membrane proteins via simulation. J Am Chem Soc 128:2697–2704
11. Born M (1920) Volumen und Hydratationswärme der Ionen. Z Phys 1:45–48
12. Still WC, Tempczyk A, Hawley RC, Hendrickson T (1990) Semianalytical treatment of solvation for molecular mechanics and dynamics. J Am Chem Soc 112:6127–6129
13. Jang S, Shin S, Pak Y (2002) Molecular dynamics study of peptides in implicit water: ab initio folding of beta-hairpin, beta-sheet, and beta beta alpha-motif. J Am Chem Soc 124:4976–4977
14. Simmerling C, Strockbine B, Roitberg AE (2002) All-atom structure prediction and folding simulations of a stable protein. J Am Chem Soc 124:11258–11259
15. Snow CD, Nguyen N, Pande VS, Gruebele M (2002) Absolute comparison of simulated and experimental protein-folding dynamics. Nature 420:102–106
16. Chowdhury S, Zhang W, Wu C, Xiong GM, Duan Y (2003) Breaking non-native hydrophobic clusters is the rate-limiting step in the folding of an alanine-based peptide. Biopolymers 68:63–75
17. Ulmschneider JP, Jorgensen WL (2004) Polypeptide folding using Monte Carlo sampling, concerted rotation, and continuum solvation. J Am Chem Soc 126:1849–1857
18. Spassov VZ, Yan L, Szalma S (2002) Introducing an implicit membrane in generalized Born/solvent accessibility continuum solvent models. J Phys Chem B 106:8726–8738
19. Tanizaki S, Feig M (2005) A generalized Born formalism for heterogeneous dielectric environments: application to the implicit modeling of biological membranes. J Chem Phys 122:124706
20. Tanizaki S, Feig M (2006) Molecular dynamics simulations of large integral membrane proteins with an implicit membrane model. J Phys Chem B 110:548–556
21. Im W, Chen JH, Brooks CL (2006) Peptide and protein folding and conformational equilibria: theoretical treatment of electrostatics and hydrogen bonding with implicit solvent models. In: Baldwin R, Baker D (eds) Peptide solvation and H-bonds. Elsevier Academic, San Diego
22. Ulmschneider JP, Ulmschneider MB, Di Nola A (2007) Monte Carlo folding of transmembrane helical peptides in an implicit generalized Born membrane. Proteins 69:297–308
23. Bu L, Im W, Brooks CL III (2007) Membrane assembly of simple helix homo-oligomers studied via molecular dynamics simulations. Biophys J 92:854–863
24. Lee J, Im W (2007) Implementation and application of helix-helix distance and crossing angle restraint potentials. J Comput Chem 28:669–680
25. Ulmschneider MB, Ulmschneider JP, Sansom MSP, Di Nola A (2007) A generalized Born implicit membrane representation compared to experimental insertion free energies. Biophys J 92:2338–2349
26. Im W, Feig M, Brooks CL III (2003) An implicit membrane generalized born theory for the study of structure, stability, and interactions of membrane proteins. Biophys J 85: 2900–2918

27. Im W, Brooks CL III (2004) De novo folding of membrane proteins: an exploration of the structure and NMR properties of the fd coat protein. J Mol Biol 337:513–519
28. Im W, Brooks CL (2005) Interfacial folding and membrane insertion of designed peptides studied by molecular dynamics simulations. Proc Natl Acad Sci U S A 102:6771–6776
29. Ulmschneider MB, Sansom MS, Di Nola A (2005) Properties of integral membrane protein structures: derivation of an implicit membrane potential. Proteins Struct Funct Genet 59:252–265
30. Hessa T, Kim H, Bihlmaier K, Lundin C, Boekel J, Andersson H, Nilsson I, White SH, Von Heijne G (2005) Recognition of transmembrane helices by the endoplasmic reticulum translocon. Nature 433:377–381
31. Hessa T, White SH, Von Heijne G (2005) Membrane insertion of a potassium-channel voltage sensor. Science 307:1427
32. Totrov M (2004) Accurate and efficient generalized born model based on solvent accessibility: derivation and application for LogP octanol/water prediction and flexible peptide docking. J Comput Chem 25:609–619
33. Shih AY, Arkhipov A, Freddolino PL, Schulten K (2006) Coarse grained protein-lipid model with application to lipoprotein particles. J Phys Chem B 110:3674–3684
34. Bond PJ, Holyoake J, Ivetac A, Khalid S, Sansom MSP (2007) Coarse-grained molecular dynamics simulations of membrane proteins and peptides. J Struct Biol 157:593–605
35. Qiu D, Shenkin PS, Hollinger FP, Still WC (1997) The GB/SA continuum model for solvation. A fast analytical method for the calculation of approximate Born radii. J Phys Chem A 101:3005–3014
36. Jorgensen WL, Ulmschneider JP, Tirado-Rives J (2004) Free energies of hydration from a generalized Born model and an ALL-atom force field. J Phys Chem B 108:16264–16270
37. Parsegian A (1969) Energy of an ion crossing a low dielectric membrane: solutions to four relevant electrostatic problems. Nature 221:844–846
38. Schaefer M, Bartels C, Karplus M (1998) Solution conformations and thermodynamics of structured peptides: molecular dynamics simulation with an implicit solvation model. J Mol Biol 284:835–848
39. Radzicka A, Wolfenden R (1988) Comparing the polarities of the amino-acids—side-chain distribution coefficients between the vapor-phase, cyclohexane, 1-octanol, and neutral aqueous-solution. Biochemistry 27:1664–1670
40. Ulmschneider JP, Jorgensen WL (2003) Monte Carlo backbone sampling for polypeptides with variable bond angles and dihedral angles using concerted rotations and a Gaussian bias. J Chem Phys 118:4261–4271
41. Jorgensen WL, Maxwell DS, Tirado-Rives J (1996) Development and testing of the OPLS all-atom force field on conformational energetics and properties of organic liquids. J Am Chem Soc 118:11225–11236
42. Ulmschneider JP, Jorgensen WL (2004) Monte Carlo backbone sampling for nucleic acids using concerted rotations including variable bond angles. J Phys Chem B 108: 16883–16892
43. Ulmschneider JP, Ulmschneider MB, Di Nola A (2006) Monte Carlo vs molecular dynamics for all-atom polypeptide folding simulations. J Phys Chem B 110:16733–16742
44. Earl DJ, Deem MW (2005) Parallel tempering: theory, applications, and new perspectives. Phys Chem Chem Phys 7:3910–3916
45. Daura X, Gademann K, Jaun B, Seebach D, Van Gunsteren WF, Mark AE (1999) Peptide folding: when simulation meets experiment. Angew Chem Int Ed 38:236–240
46. Mackenzie KR, Prestegard JH, Engelman DM (1997) A transmembrane helix dimer: structure and implications. Science 276:131–133
47. Smith SO, Song D, Shekar S, Groesbeek M, Ziliox M, Aimoto S (2001) Structure of the transmembrane dimer interface of glycophorin A in membrane bilayers. Biochemistry 40:6553–6558
48. Mackenzie KR (2006) Folding and stability of alpha-helical integral membrane proteins. Chem Rev 106:1931–1977

49. Efremov RG, Nolde DE, Konshina AG, Syrtcev NP, Arseniev AS (2004) Peptides and proteins in membranes: what can we learn via computer simulations? Curr Med Chem 11:2421–2442
50. White SH, Wimley WC (1999) Membrane protein folding and stability: physical principles. Annu Rev Biophys Biomol Struct 28:319–365
51. White SH (2005) How hydrogen bonds shape membrane protein structure. Adv Protein Chem 72:157–172
52. Im W, Chen J, Brooks CL 3rd (2005) Peptide and protein folding and conformational equilibria: theoretical treatment of electrostatics and hydrogen bonding with implicit solvent models. Adv Protein Chem 72:173–98
53. Petrache HI, Grossfield A, Mackenzie KR, Engelman DM, Woolf TB (2000) Modulation of glycophorin A transmembrane helix interactions by lipid bilayers: molecular dynamics calculations. J Mol Biol 302:727–746
54. Efremov R, Volynsky P, Nolde D, Vergoten G, Arseniev A (2001) Implicit two-phase solvation model as a tool to assess conformation and energetics of proteins in membrane-mimetic media. Theor Chem Acc 106:48–54
55. Williams S, Causgrove TP, Gilmanshin R, Fang KS, Callender RH, Woodruff WH, Dyer RB (1996) Fast events in protein folding: helix melting and formation in a small peptide. Biochemistry 35:691–697
56. Clarke DT, Doig AJ, Stapley BJ, Jones GR (1999) The alpha-helix folds on the millisecond time scale. Proc Natl Acad Sci U S A 96:7232–7237
57. Paci E, Cavalli A, Vendruscolo M, Caflisch A (2003) Analysis of the distributed computing approach applied to the folding of a small beta peptide. Proc Natl Acad Sci U S A 100:8217–8222
58. Snow CD, Qiu LL, Du DG, Gai F, Hagen SJ, Pande VS (2004) Trp zipper folding kinetics by molecular dynamics and temperature-jump spectroscopy. Proc Natl Acad Sci U S A 101:4077–4082
59. Opella SJ, Marassi FM, Gesell JJ, Valente AP, Kim Y, Oblatt-Montal M, Montal M (1999) Structures of the M2 channel-lining segments from nicotinic acetylcholine and NMDA receptors by NMR spectroscopy. Nat Struct Biol 6:374–379
60. Law RJ, Forrest LR, Ranatunga KM, La Rocca P, Tieleman DP, Sansom MS (2000) Structure and dynamics of the pore-lining helix of the nicotinic receptor: MD simulations in water, lipid bilayers, and transbilayer bundles. Proteins Struct Funct Genet 39:47–55
61. Law RJ, Tieleman DP, Sansom MSP (2003) Pores formed by the nicotinic receptor M2 delta peptide: a molecular dynamics simulation study. Biophys J 84:14–27
62. Milik M, Skolnick J (1993) Insertion of peptide chains into lipid membranes: an off-lattice Monte Carlo dynamics model. Proteins Struct Funct Genet 15:10–25
63. Maddox MW, Longo ML (2002) A Monte Carlo study of peptide insertion into lipid bilayers: equilibrium conformations and insertion mechanisms. Biophys J 82:244–263
64. Kessel A, Shental-Bechor D, Haliloglu T, Ben-Tal N (2003) Interactions of hydrophobic peptides with lipid bilayers: Monte Carlo simulations with M2delta. Biophys J 85:3431–3444
65. Kessel A, Haliloglu T, Ben-Tal N (2003) Interactions of the M2delta segment of the acetylcholine receptor with lipid bilayers: a continuum-solvent model study. Biophys J 85:3687–3695

Multi-scale Approaches to Dynamical Transmission of Protein Allostery

Philip D. Townsend, Thomas L. Rodgers, Ehmke Pohl, Mark R. Wilson, Martin J. Cann, and Tom C.B. McLeish

Abstract We review the idea that allosteric interactions can be transmitted not by structural switching but by the more subtle route of modulation of the amplitude of thermally-activated global dynamical modes in allosteric proteins. The effect is naturally addressed and explored through coarse-grained models of protein dynamics, but can be linked to atomistic models of substrate binding at the fine scale, and to themodynamic free energies at the macroscopic scale. A remarkable specificity at the residue level emerges: allosteric proteins possess a set of 'control sites' whose modification by single-point mutation may alter allosteric free-energies non-perturbatively.

Keywords Allostery • Protein physics • Dynamics • Elastic network • Thermal fluctuations • Biophysics • Molecular simulation • Elastic network models • CAP protein • Normal modes • Conformational changes • Multiscale modelling

---

P.D. Townsend • M.J. Cann
Biophysical Sciences Institute, Durham University, Durham, UK

School of Biological and Biomedical Sciences, Durham University, Durham, UK

T.L. Rodgers
School of Chemical Engineering and Analytical Sciences, University of Manchester, Manchester, UK

Biophysical Sciences Institute, Durham University, Durham, UK

Department of Chemistry, Durham University, Durham, UK

E. Pohl
School of Biological and Biomedical Sciences, Durham University, Durham, UK

Biophysical Sciences Institute, Durham University, Durham, UK

Department of Chemistry, Durham University, Durham, UK

M.R. Wilson
Biophysical Sciences Institute, Durham University, Durham, UK

Department of Chemistry, Durham University, Durham, UK

T.C.B. McLeish (✉)
Department of Physics, Durham University, Durham, UK

Biophysical Sciences Institute, Durham University, Durham, UK

Department of Chemistry, Durham University, Durham, UK
e-mail: t.c.b.mcleish@durham.ac.uk

L. Olivares-Quiroz et al. (eds.), *Physical Biology of Proteins and Peptides*,
DOI 10.1007/978-3-319-21687-4_8

Introduction

Allostery describes a fundamental mechanism of both normal and pathological molecular cellular processes [1] by which substrate binding at one site on a protein affects the binding strength at another, distant, site. Early models of allosteric co-operativity invoked binding-induced switching between distinct protein conformers. In the 'symmetry model' (MWC: Monod-Wyman-Changeux model), an equilibrium between two conformational states is perturbed by ligand binding such that all subunits in the oligomeric protein change state in a concerted fashion [2]. In the 'sequential model' (KNF: Koshland-Nemethy-Filmer model) a requirement for conformational symmetry is replaced with a strict application of 'induced-fit' binding of ligand with distinct protein conformers in the bound and unbound states [3]. These influential models facilitated the development of a view of allostery dominated by a view that structural change was the underlying mechanism.

Less familiar has been the development of and discussion around an alternative mechanism which does not require structural switching, but instead exploits the fact that irrespective of any structural change, substrate binding also modifies the amplitudes of thermal fluctuations *around* the mean structure. Since such a restriction of random motion constitutes in turn a change in entropy, the full allosteric free energy may contain components that arise purely by this route [4]. A feature of this mechanism is that to contribute to non-local allosteric interaction, the longer-wavelength low-frequency 'global' modes of motion are recruited, rather than the higher-frequency and more local motions (such as side-group oscillation). Successful models of protein dynamics that capture the effect are therefore coarse-grained rather than atomically-resolved. Specific models of particular protein systems at various degrees of course-graining have been constructed [5–7] which show that the orders of magnitude of real allosteric free energies can be generated by such restriction of dynamical correlations alone. Similar challenges are presented to experiments, which need to identify the large-scale dynamical changes at many points on a protein on substrate-binding, and to connect them to thermodynamics. NMR [8, 9], X-ray analysis of B-factors and isothermal calorimetry [10] have been deployed in comparison with modelling.

Protein Dynamics in Allostery

The theoretical study of Cooper and Dryden in [4] demonstrated that allosteric interaction free energies of the order of several kJ mol^{-1} could be achieved through ligand induced changes in thermally-activated protein dynamics. Specifically, the dynamic allostery described is achieved primarily through altered entropy, arising from the ligand-modulated amplitude of protein thermal motion. The idea implies that proteins have evolved to take functional advantage of not only the mean conformation, but also the inherent thermal fluctuations about this mean. It was found that *root mean square* shifts of only 1 % per atom, when summed over an entire protein, could permit changes to the allosteric free energies purely through alterations in the entropy of fluctuations in internal conformations [4]. Following from this theoretical analysis, recent developments in nuclear magnetic resonance (NMR) spectroscopy has revolutionised our experimental capacity to study the role of protein structure, dynamics and thermodynamics in allosteric systems [11]. A particularly valuable development has been the use of relaxation NMR methodologies that use fast internal protein dynamics as a proxy for changes in conformation entropy on ligand binding [12]. It is become clear, in consequence of both experimental and theoretical work, that dynamics has a crucial role to play in allostery.

A Coarse-Grained Picture

It is instructive to develop coarse-grained models of dynamic allostery for several reasons: (1) they serve to illuminate the general principle by which this subtle mechanism operates; (2) they allow exact calculation of allosteric free energies in the corresponding simplified cases; (3) the consequence of changing the small number of fundamental parameters of the models may be visualised in a straightforward way as response-surfaces. We demonstrate these three principles in this section in the context of a simple example.

In general, a large molecule such as a globular protein has many degrees of freedom that may be coupled in complex ways. However, if the couplings are all harmonic (quadratic in the coordinates) then the Gaussian partition function and free energy of the single mode generalises in a straightforward manner. In the general case, the full Hamiltonian is

$$\mathrm{H} = \frac{1}{2}\mathbf{p}^{T}\mathrm{M}^{-1}\mathbf{p} + \mathbf{x}\mathrm{K}\mathbf{x} \tag{1}$$

where $\mathbf{p}$ is the vector of the momenta and $\mathbf{x}$ that of coordinates. M and K are the (diagonal) mass matrix and (symmetric) spring coupling (Hessian) matrix respectively. Models of just this type are generated as realistic representations of proteins through the Elastic Network Model (ENM) approach [13]. In the case of just a single harmonic degree of freedom, the standard expression for the partition function Z is just an integral over the Gaussian fluctuations of the mode, controlled by the spring constant K. When taking the free energy *via* $F = -k_B T \log Z$, the value becomes logarithmically dependent on the spring constant as $F = k_B T \log K$ with other constant terms. In the case of a more complex dynamical object, yet still harmonic, this entropic free energy can be decomposed as a sum over the contributions from the normal models, becoming now a logarithmic function of a matrix determinant. The allosteric free energy is now, in turn, a logarithmic function of the determinants of the many-body harmonic system with zero, one and two effectors bound.

$$G = \frac{1}{2}k_B T \ln \frac{|\mathrm{K}_0||\mathrm{K}_2|}{|\mathrm{K}_1|^2} \tag{2}$$

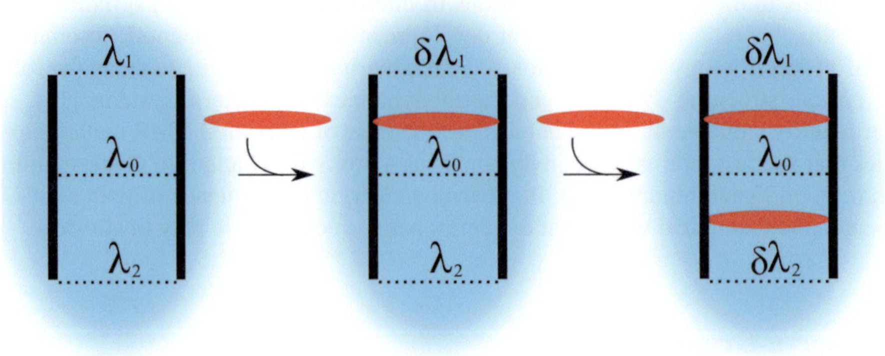

Fig. 1 A highly coarse-grained model of a homodimer, in which only relative motion of the two domains is modelled. They are connected across their interface by three flexible springs. These are modulated locally by binding to substrates, changing their effective spring constants, λ, by a factor δ

Here K_1 is the Hessian for 1 effector bound, and likewise for 0 and 2 bound. We illustrate the richness that emerges from the mechanism in a highly simplified case of a coarse-grained model for a homodimer. The two domains are assumed to be rigid, but attached at three points by harmonic bonds. Substrate binding at the first and third of these may modify the strength of the bonds, so generating an allosteric interaction (Fig. 1). These are modulated locally by binding to substrates, changing their effective spring constants, λ, by a factor δ. There are two degrees of freedom in this simple system: as well as the displacement coordinate between the two rigid rods, x, there is also now an angular displacement between the two rods away from their parallel equilibrium, θ. The vector of generalised coordinates, the potential energy matrix and the allosteric free energy of the system has the structure of (3):

$$
\begin{aligned}
\mathbf{x} &= \begin{pmatrix} x \\ \theta \end{pmatrix} \\
\mathrm{K} &= \begin{pmatrix} \lambda_1+\lambda_2+\lambda_0 & \lambda_1-\lambda_2 \\ \lambda_1-\lambda_2 & \lambda_1+\lambda_2 \end{pmatrix} \\
G &= \frac{1}{2}k_B T \ln\left(\frac{4(1+\delta)(2+2\delta+\lambda_0)(2+\lambda_0)}{\left[4(1+\delta)+\lambda_0(2+\delta)\right]^2} \right) \\
&= 0 \quad \text{when } \lambda_0 = 0
\end{aligned}
\tag{3}
$$

Here the spring constants at the binding sites λ_1 and λ_2, are set (without loss of generality) to 1 and are enhanced by a dimensionless increment δ on substrate binding. The dimensionless hinge spring constant λ_0 becomes a second parameter in the now two dimensional landscape for the allosteric landscape $\Delta\Delta G$. Remarkably, however, even if a large value of the substrate effect δ is chosen yet the central hinge stiffness is set to zero, the allosteric effect vanishes. This might be thought of as arising from the extreme symmetry of the simple system. However, the same symmetry of the molecule pertains in the case of non-zero values of λ_0. Rather, the cancellation arises because of the pure antisymmetry of the two (only) modes of this system. Although each carries an allosteric effect, they are exactly opposite. But their antisymmetry is broken as soon as the central spring assumes a non-zero value. This is the simplest case of a general result [14] that inhomogeneities in elasticity within a dynamically allosteric system are necessary to transmit a non-zero signal. In this light, the strong variations in density within globular proteins constitute a functional requirement that fluctuation-dynamical pathways for allostery exist.

Normal Mode Analysis with Example of Catabolite Activator Protein

The allosteric mechanism proposed here is that large scale motion delocalised across the protein via the normal modes of deformation, can carry the allosteric signal without a requirement for structural change [5–7]. At first glance this appears to be a radically different mechanism from a number of hypotheses that have been presented to explain the mechanism by which dynamic fluctuations are communicated between allosterically coupled sites in various protein systems. One hypothesis is that ligand binding in an allosterically regulated system changes the ensemble of the protein structures, incorporating the allosteric sites, and therefore the free energy of binding [15–17]. Yet, providing that the fluctuations around the mean structure sample an effective continuum of states (this is just the classical limit, appropriate for the larger-scale and slower modes of protein dynamics), this 'ensemble of states' mechanism is equivalent to the 'fluctuation amplitude modulation' mechanism. A third alternative proposal has discussed physically connected pathways of activated or repressed dynamics between allosteric sites [18–20]. This may actually constitute a distinct mechanism, however in some cases the structure of the combined normal modes may appear approximately to constitute a spatial pathway. This is a question that will require further investigation.

The normal modes describe the different harmonic vibrational oscillations in a protein around a mean minimum energy structure. Normal mode analysis (NMA) represents a major simulation technique to study large-scale motions in proteins [21, 22]. In proteins, the low frequency modes are typically associated with function through collected motion while the higher frequency modes are spatially localized due to dynamic disorder, associated with protein structural disorder [13]. NMA of an all atom protein structure represents a significant computational problem, particularly due to the long time-scales required for a full analysis of the low frequency modes. One approach to circumvent this problem is the use of elastic network models (ENM) in which the total number of atoms in the protein is reduced to only Cα atoms connected by a network of elastic Hookean springs [23]. ENM based models reproduce low-frequency modes that compare well experimentally derived data [24, 25]. Coarse-grained models can therefore provide a reasonably accurate calculation of dynamics over the long time scales required.

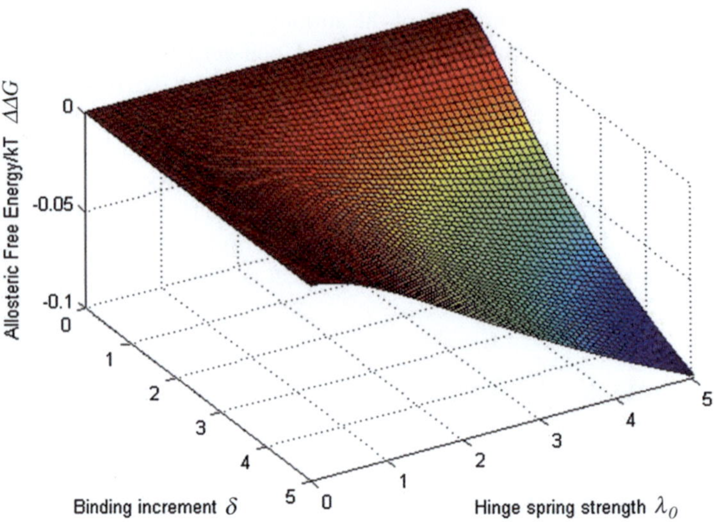

Fig. 2 The 2-dimensional response surface for the allosteric free energy in terms of the two independent parameters of the model (incremental spring strength change on effector binding and relative strength of the central tethering potential) units of *kT*

We have applied the ENM level of coarse-graining, within the framework of modulated fluctuation, to the strongly negatively allosteric regulation of a truncated variant of the Catabolite Activator Protein (CAP) [26]. CAP is a 210 amino acid homodimeric transcription factor of *Escherichia coli* that binds cAMP generated by adenylyl cyclase in response to the phosphorylated form of Enzyme IIA^{Glc} [27]. CAP consists of two domains per monomer. The N-terminal 138 amino acids represent the ligand binding domain while the C-terminal portion of the molecule represents the DNA binding domain (Fig. 3). A range of NMR studies have shown that conformational entropy via altered backbone and side-chain dynamics is linked to positive allostery between ligand and DNA binding where cAMP binding promotes the interaction with DNA [8, 9]. Negative allostery also exists in native CAP between the cAMP binding sites and is of the order of $K_2/K_1 = 1.7$ (where K_1 and K_2 represent the dissociation constants for the first and second cAMP binding events) ([26]—and for a discussion on the functional role of negative allostery in CAP see [10]).

The allosteric free energy for binding the second molecule of cAMP in the calculated ENM was remarkably similar to experimentally obtained values indicating that the normal modes could successfully reproduce allostery in CAP [10, 28]. The value of K_2/K_1 (where K_2/K_1 represents the ratio of the second to first dissociation constants for cAMP) obtained for independent wild type ENMs ($K_2/K_1 = 1.35$) was similar to that obtained experimentally through isothermal titration calorimetry ($K_2/K_1 = 1.68 \pm 0.04$; S.E.M., $n = 32$). The ENM therefore captures negative allostery in CAP. Analysis of the cumulative protein motion in the CAP ENM reveals that the first 75 modes represent approximately 90 % of the total motion within the model (Fig. 2). The allostery captured by the ENM is therefore represented by the lower

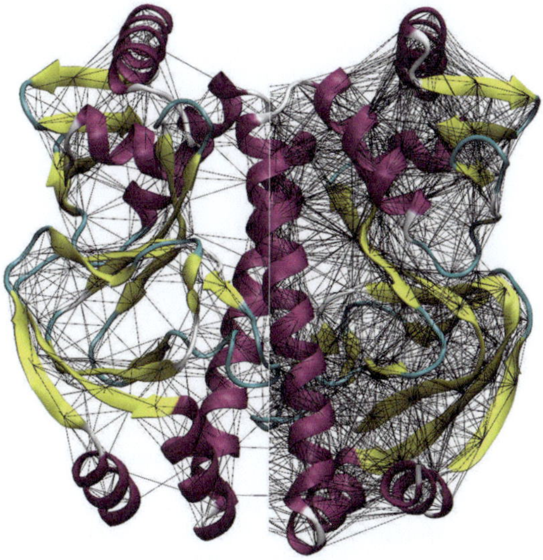

Fig. 3 ENM models for the CAP homodimer using 8 Å (*left half*) and 12 Å (*right half*) cut-offs for the insertion of Cα-coupling harmonic springs, superposed on a ribbon diagram of the crystal structure of CAP determined at 1.48 Å resolution (PDB 4HZF) showing the secondary and tertiary structures of the CAP homodimer

frequency modes that represent the majority of the motion in CAP. Further, an analysis of just the first 25 modes, representing 78 % of the total protein motion, within 16 ENMs derived from independent X-ray CAP structures is still able to capture a significant degree of negative allostery ($K_2/K_1 = 1.13 \pm 0.01$).

A Mutational Map for Dynamic Control of Allostery in CAP

A strong check on the validity of coarse-grained fluctuation models of allostery is to investigate their predictions when single sites are mutated so that the local elasticity in the neighbourhood of those sites is modified. As the ENM is a computationally efficient method for calculating $\Delta\Delta G$, it was possible to vary the elastic force constant at every Cα atom in turn and calculate the effect on allosteric connectivity. At each site the spring constant (k_R/k; corresponds to $k_{\text{amino acid number}}$/relative spring strength and $k_R/k = 1$ throughout the wild type ENM) could be increased (to represent a strengthened side chain interaction) or decreased (to represent a weakened side chain interaction) and $\Delta\Delta G$ calculated. The resulting allosteric map, a two-dimensional colour coded surface that related k_R/k for every amino acid to K_2/K_1, could reveal local features of the protein that exerted control over allostery (Fig. 4). A cursory examination of Fig. 4 reveals that residues that exert an influence over allostery in CAP are distributed over the entire CAP monomer but also with a tendency to form clusters (e.g., at the dimer interface). Furthermore, within these coarse-grained clusters is a highly fine-grained structure that differentiates between the effect of individual residues. In consequence, there exists a small set of 'allosteric control residues', whose modification by mutation is predicted to alter the allosteric

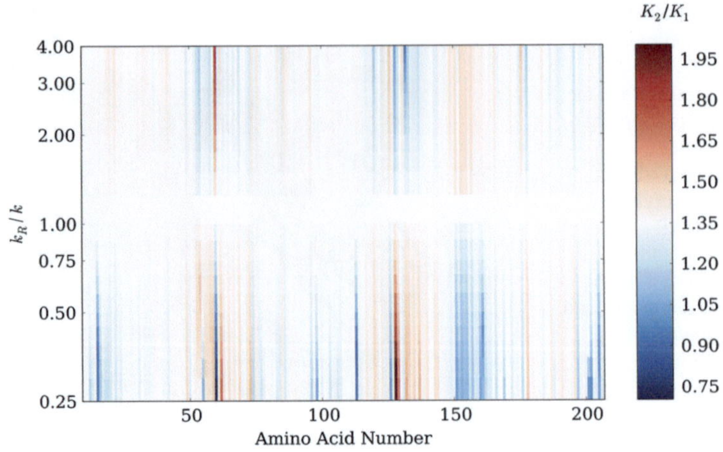

Fig. 4 Global map for an ENM for CAP plotting amino acid number for the CAP monomer and dimensionless change in k_R/k. The colour chart represents changes in the ratio of the second to first dissociation constants for cAMP. White corresponds to values of K_2/K_1 predicted by the wild type ENM. *Red* corresponds to increased values of K_2/K_1 (increased negative co-operativity) and *blue* corresponds to decreased values of K_2/K_1 (decreased negative co-operativity and positive co-operativity). Figure taken from [10]

interaction between the cAMP binding sites significantly. This gives a 'second order' "control at a distance" whereby modifications in the ENM modulate allostery through changes in binding entropy by a global change in the normal modes, itself controlled at a third site. These sensitive control sites are identified in a real-space representation of the CAP dimer in Fig. 5.

A key feature of the ENM for CAP and the insight obtained into the role of low frequency global motions in allostery, is that it enables clear predictions for experimental validation. The global allosteric control map for the ENM (Fig. 4) enabled the selection of allosteric control residues for further testing through the introduction of mutations into the CAP protein and an assessment of allostery by isothermal calorimetry (ITC). Isothermal calorimetry is a standard experimental tool for the experimental assessment of the thermodynamics of protein ligand interactions that directly calculates K_a, ΔH, and binding stoichiometry and subsequently ΔG and ΔS from the Gibb's equation [29]. Rodgers et al. selected amino acids at the CAP dimer interface (V132), and on a surface loop distant from both the cAMP binding site and the dimer interface (H160) that were predicted to modulate allostery on side chain mutation. An additional site (V140) was selected as it was predicted to have no influence on allosteric control. It was observed that mutations predicted through the ENM to increase the extent of negative co-operativity (V132A), decrease the extent of negative co-operativity (H160L), or switch allostery in CAP from negative to positive co-operativity (V132L), all behaved as expected when analysed experimentally. In each case there was a semi-quantitative relationship between the degree of change in K_2/K_1 predicted by the ENM (Fig. 5)

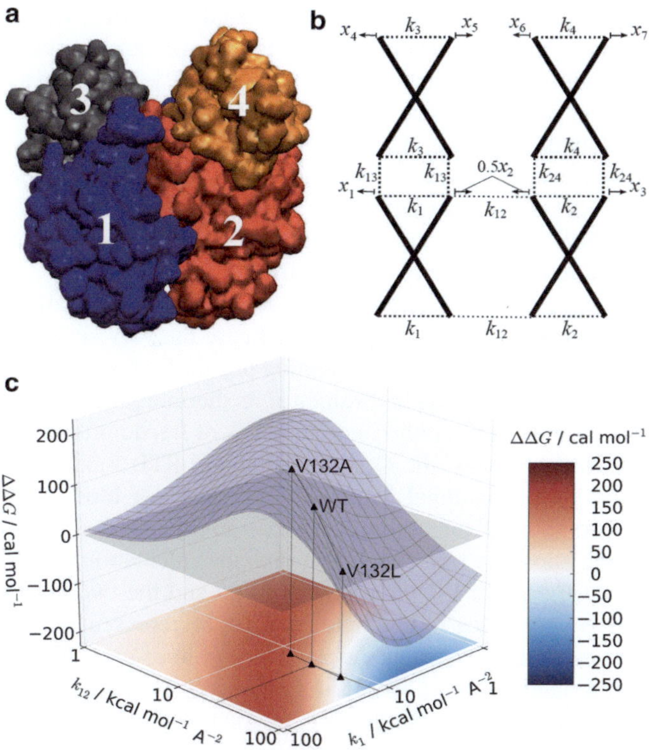

Fig. 5 A super-coarse-grained models for CAP. (**a**) The elastic block representation for CAP that emerges from constraining all residues whose relative spatial fluctuations are less than 3 Å into a single rigid domain. (**b**) The corresponding model in which each block is accorded a single internal mode. (**c**) Plot of allosteric free energy in CAP as a function of the reduced two-dimensional landscape of the effective internal (k_1) and coupling (k_{12}) elastic strengths of the super-coarse-grained model. Figure taken from [10]

and that observed experimentally. A subset of amino acids in CAP is therefore predicted to exert control over allostery and the data for several of these are experimentally supported.

Global Protein Properties for Dynamic Control of Allostery

The theoretical analysis of allostery in CAP discussed here is suggestive of general underlying biophysical principles that permit allostery. In particular it strongly suggests that the central contribution to fluctuation-modulated allostery emerges at the global, coarse-grained, levels of protein structure. We have seen that the step from fully atomistic resolution to that of the ENM correlates with experimentally observed thermodynamics across a range of mutated structures. This approach can be carried further by coarse-graining CAP into a rotational-translational block (RTB) representation [30]. In such a representation CAP is modelled as two coarse-grained blocks per monomer (one for the cAMP-binding domain and one for the DNA-binding domain) (Fig. 5a). Each individual block is rigid in this model, with internal residues fixed in relation to one another but with an integrated coupling strength. The blocks are connected to their neighbours by the remaining springs of the ENM (Fig. 5b). The rigid blocks are permitted one internal breathing motion and are permitted to move relative to each other.

The RTB model can be understood in one sense as a generalisation of the 'scissor' model introduced earlier. However, in that case the very simple structure was introduced purely for the purpose of illustrating physically how the effect arises and to calculate the broad parameters and values of its thermodynamics. An RTB model of a multi-domain protein has rather more degrees of freedom than the generalised scissor model. In the case of the CAP, four-domain RTB model, there are three translational and three rotational degrees of freedom per block, reduced by the translation and rotation of the entire complex—so nine degrees of freedom. The calculation of the allosteric free energy requires the calculation of the determinant of a 9×9 matrix in each of the three cases of apo and the two holo forms of the dimer. But this is very quick to compute, and can be represented as a function of reduced sets of parameters, such as the internal and inter-monomer coupling constants, generating landscapes of allosteric free energy parameterised by such biophysically significant and coarse-grained interaction values (k_1 and k_{12} in the simplified diagram for the RTB structure in Fig. 5b).

Remarkably, negative allostery can even be captured in such a super-coarse-grained model system for CAP ($K_2/K_1=1.08$) [14, 31]. The actual position of wild type CAP, or its variants, represent single points on a larger allosteric free energy landscape and the position of any CAP variant is predominantly determined at this level of coarse-graining by changes to the coupling strength within the ligand binding domain (k_1) and between monomers (k_{12}) (Fig. 5c). The reduction to this coarse-grained landscape shows that the wild-type CAP protein is situated at a region of its parameter space close to, but not exactly on, a 'ridge' of maximal negative allosteric response. The response can be made more negative still by increasing k_1 and decreasing k_{12} from wild type values.

The description of negative allostery in a super-coarse grained model for CAP raises an even more fundamental question. What are the required properties of a protein that can permit low frequency dynamics controlled allostery to arise? McLeish and co-authors performed a series of calculations to exemplify the fundamental properties of fluctuation-induced allostery [14]. The highly coarse-grained "toy" systems analysed revealed that a key requirement is an inhomogeneous elastic density in the system. In a protein, such elastic inhomogeneity can arise through both relatively rigid and less rigid regions within the internal environment of the protein and also through the boundary geometry of the folded protein. This is observed at a simple level in the super-coarse-grained model for CAP, but note that the density of Cα-Cα springs in the construction of the ENM model direct from PDB structures (see Fig. 3) motivates such a rigid block construction naturally.

Significantly, dynamically induced allostery that operates through modulation of the normal modes has the potential for being both positive or negative in sign so long as there are at least two coupled degrees of freedom within the system [31]. This theory therefore suggests that the requirements for an appropriately tuned internal elastic architecture that supports the functional requirements for allostery in a protein can co-exist with the structural requirements that underpin other aspects of protein function.

Conclusions

A mechanism of allosteric interaction by which substrate or effector binding modulates the amplitude of thermal fluctuations around a mean structure, rather than activates a mean structural change, is now demonstrated both theoretically and experimentally. The CAP protein of *E. coli* has served as an exemplar to demonstrate how such modulation of fluctuation, broken down into the normal modes of the protein dynamics can underpin allosteric co-operativity. Analysis of the theoretical framework for allostery in this model has allowed both the rational engineering of allostery and the identification of fundamental protein properties that permit its control in magnitude and sign. CAP represents a special case of dynamic allostery where the mechanism is almost entirely entropically driven. Challenges for the future include both the assessment of the contribution of fluctuation modulation to allosteric systems that are also characterised by conformational change, and the identification of protein designs that optimise fluctuation-induced allostery.

References

1. Motlagh HN, Wrabl JO, Li J, Hilser VJ (2014) The ensemble nature of allostery. Nature 508:331–339. doi:10.1038/nature13001
2. Monod J, Wyman J, Changeux JP (1965) On the nature of allosteric transitions—a plausible model. J Mol Biol 12:88–118
3. Koshland DE Jr, Nemethy G, Filmer D (1966) Comparison of experimental binding data and theoretical models in proteins containing subunits. Biochemistry 5:365–385
4. Cooper A, Dryden DT (1984) Allostery without conformational change. A plausible model. Eur Biophys J 11:103–109
5. Hawkins RJ, McLeish TC (2004) Coarse-grained model of entropic allostery. Phys Rev Lett 93:098104
6. Hawkins RJ, McLeish TCB (2006) Coupling of global and local vibrational modes in dynamic allostery of proteins. Biophys J 91:2055–2062
7. Hawkins RJ, McLeish TCB (2006) Dynamic allostery of protein alpha helical coiled-coils. J R Soc Interface 3:125–138
8. Tzeng SR, Kalodimos CG (2009) Dynamic activation of an allosteric regulatory protein. Nature 462:368–372. doi:10.1038/nature08560
9. Tzeng SR, Kalodimos CG (2012) Protein activity regulation by conformational entropy. Nature 488:236–240. doi:10.1038/nature11271

10. Rodgers TL et al (2013) Modulation of global low-frequency motions underlies allosteric regulation: demonstration in CRP/FNR family transcription factors. PLoS Biol 11, e1001651. doi:10.1371/journal.pbio.1001651
11. Manley G, Rivalta I, Loria JP (2013) Solution NMR and computational methods for understanding protein allostery. J Phys Chem B 117:3063–3073. doi:10.1021/jp312576v
12. Wand AJ (2013) The dark energy of proteins comes to light: conformational entropy and its role in protein function revealed by NMR relaxation. Curr Opin Struct Biol 23:75–81. doi:10.1016/j.sbi.2012.11.005
13. Bahar I, Atilgan AR, Demirel MC, Erman B (1998) Vibrational dynamics of folded proteins: significance of slow and fast motions in relation to function and stability. Phys Rev Lett 80:2733–2736
14. McLeish TC, Rodgers TL, Wilson MR (2013) Allostery without conformation change: modelling protein dynamics at multiple scales. Phys Biol 10:056004. doi:10.1088/1478-3975/10/5/056004
15. Hilser VJ, Wrabl JO, Motlagh HN (2012) Structural and energetic basis of allostery. Annu Rev Biophys 41:585–609. doi:10.1146/annurev-biophys-050511-102319
16. Liu T, Whitten ST, Hilser VJ (2007) Functional residues serve a dominant role in mediating the cooperativity of the protein ensemble. Proc Natl Acad Sci U S A 104:4347–4352
17. Motlagh HN, Hilser VJ (2012) Agonism/antagonism switching in allosteric ensembles. Proc Natl Acad Sci U S A 109:4134–4139
18. Reynolds KA, McLaughlin RN, Ranganathan R (2011) Hot spots for allosteric regulation on protein surfaces. Cell 147:1564–1575. doi:10.1016/j.cell.2011.10.049
19. Zhuravleva A, Clerico EM, Gierasch LM (2012) An interdomain energetic tug-of-war creates the allosterically active state in Hsp70 Molecular chaperones. Cell 151:1296–1307. doi:10.1016/j.cell.2012.11.002
20. Zhuravleva A, Gierasch LM (2011) Allosteric signal transmission in the nucleotide-binding domain of 70-kDa heat shock protein (Hsp70) molecular chaperones. Proc Natl Acad Sci U S A 108:6987–6992. doi:10.1073/pnas.1014448108
21. Brooks B, Karplus M (1983) Harmonic dynamics of proteins: normal modes and fluctuations in bovine pancreatic trypsin inhibitor. Proc Natl Acad Sci U S A 80:6571–6575
22. Go N, Noguti T, Nishikawa T (1983) Dynamics of a small globular protein in terms of low-frequency vibrational modes. Proc Natl Acad Sci U S A 80:3696–3700
23. Tirion MM (1996) Large amplitude elastic motions in proteins from a single-parameter, atomic analysis. Phys Rev Lett 77:1905–1908
24. Delarue M, Sanejouand YH (2002) Simplified normal mode analysis of conformational transitions in DNA-dependent polymerases: the elastic network model. J Mol Biol 320:1011–1024
25. Valadie H, Lacapcre JJ, Sanejouand YH, Etchebest C (2003) Dynamical properties of the MscL of Escherichia coli: a normal mode analysis. J Mol Biol 332:657–674
26. Popovych N, Sun S, Ebright RH, Kalodimos CG (2006) Dynamically driven protein allostery. Nat Struct Mol Biol 13:831–838
27. Gorke B, Stulke J (2008) Carbon catabolite repression in bacteria: many ways to make the most out of nutrients. Nat Rev Microbiol 6:613–624. doi:10.1038/nrmicro1932
28. Rodgers TL, Burnell D, Townsend PD, Pohl E, Cann MJ, Wilson MR, McLeish TC (2013) ΔΔPT: a comprehensive toolbox for the analysis of protein motion. BMC Bioinf 14:183. doi:10.1186/1471-2105-14-183
29. Brown A (2009) Analysis of cooperativity by isothermal titration calorimetry. Int J Mol Sci 10:3457–3477. doi:10.3390/ijms10083457
30. Durand P, Trinquier G, Sanejouand YH (1994) New approach for determining low-frequency normal-modes in macromolecules. Biopolymers 34:759–771
31. Toncrova H, McLeish TCB (2010) Substrate-modulated thermal fluctuations affect long-range allosteric signaling in protein homodimers: exemplified in CAP. Biophys J 98:2317–2326

Dynamics of Membrane Proteins and Lipid Bilayers

Héctor Eduardo Jardón-Valadez

Abstract Membrane proteins perform relevant physiological functions by means of intricate conformational changes in a hydrophobic environment. Lipid bilayers and embedded proteins, therefore, play a functional role in biomembranes, where the interplay of interactions keeps a delicate balance between cell barriers and selective transducers, transporters, pores, channels, etc. Molecular dynamics and experimental methods (e.g. X-ray diffraction, neutron scattering, nuclear magnetic resonance, infrared spectroscopy, dielectric relaxation spectroscopy, among others) encompass a set of tools to determine the relevant properties that make biomembranes so efficient for preserving life. In this chapter, I provide a perspective on studies that combine experimental methods and molecular dynamics approaches to decipher couplings of membrane proteins and lipid bilayers.

Keywords Molecular dynamics • Force fields • Neutron scattering length density • Lipid bilayers • Membrane proteins • Mean square displacement • Mean square fluctuations • Diffusion coefficient

Introduction

Membrane proteins (MP) perform a great deal of tasks for cell signaling, transport, sensing, and gating, among others [1]. MP represent 75 % of the total mass in the inner membrane of mitochondria [2]. About 30 % of the genome in microorganisms encodes for MP [3]. On the other hand, the hydrophobic core of lipid bilayers (LB) forms a barrier that preserves and protects the intracellular environment.

MP and LB are the constitutive components of biomembranes, where they may aggregate into lipid rafts—defined as microdomains of cholesterol, sphingolipids, and proteins that are involved in important functions such as signaling and trafficking

H.E. Jardón-Valadez (✉)
Departamento de Recursos de la Tierra, Autonomous Metropolitan University (Lerma), Av Hidalgo Poniente No. 46 Col. La Estación, Lerma de Villada, Estado de México 52006, México
e-mail: h.jardon@correo.ler.uam.mx

L. Olivares-Quiroz et al. (eds.), *Physical Biology of Proteins and Peptides*,
DOI 10.1007/978-3-319-21687-4_9

[4]. The complexity in biomembranes emerges as a result of its variable lipid and protein composition, which depends, for instance, on the tissue type, cellular organelle, species, and even nutrients uptake or disease conditions [5]. Moreover, at physiological conditions, biomembranes are flexible and dynamic, described as two-dimensional liquid-crystal fluid (L_α) phase [5]. The molecular assembly of MP and LB can be investigated by experimental and computational approaches for model bilayers, which provide valuable insights into the understanding of biomembranes.

This chapter is organized into three parts: first I describe the structure in a model LB, specifically the mass distribution projected along the bilayer normal; then I discuss some characteristics of the atypical lateral diffusion of lipid molecules in bilayers, that emerges as a result of its complex dynamics, self-organization, and heterogeneity; finally, I describe some of the parameters obtained by MD simulations and experimental methods that capture the lipid-protein-water dynamical couplings at comparable time scales. I conclude with a perspective on how molecular dynamics simulations contribute to the understanding of MP and LB structure and dynamics.

Lipid Bilayers

At physiological conditions LB are two-dimensional fluids, identified as L_α liquid-crystal phase [5, 6]. Along the bilayer normal, LB show a fairly well defined structure, which consist of the hydrophobic and aqueous domains, and the interfaces between them (Fig. 1). Since the pioneering studies of neutron scattering and X-ray joint refinements, devised to determine the complete structure of LB of 1,2-dioleoyl-sn-glycero-3-phosphocholine (DOPC), the mass distribution across the LB is well characterized [7]. The hydrophilic core is formed by methanediyl ($-CH_2-$) and methyl ($-CH_3$) groups of the lipid tails. This region works as a barrier for polar molecules, and its length is about 50 % of the bilayer thickness. The two hydrocarbon tails of the oleic fatty acid, esterified to the glycerol moiety in DOPC, are rather flexible; the terminal methyl groups can actually bend to reach the bilayer interface (Fig. 1a). The distribution of the DOPC functional groups along the bilayer normal reveals the length of the hydrophobic core, in particular, the region sampled by the terminal methyl groups across the bilayer (Fig. 1b). Toward the aqueous media, the phosphate and carbonyl distributions across the bilayer identify the region of the aqueous interfaces (Fig. 1b). Due to the charge distribution of the phosphocholine head group, negative at the phosphate and positive at the choline, water molecules penetrate deep into the bilayer, hydrating the lipid heads. Therefore, the heterogeneity along full length of the bilayer, display a continuum transition from the hydrophobic core to the polar aqueous media [8].

Molecular dynamics (MD) simulations using classical force fields are able to capture the mass distribution across the LB. In fact, a validation strategy for classical force fields has been proposed by comparing density profiles obtained by neutron diffraction experiments and those calculated by MD simulation [9]. Following

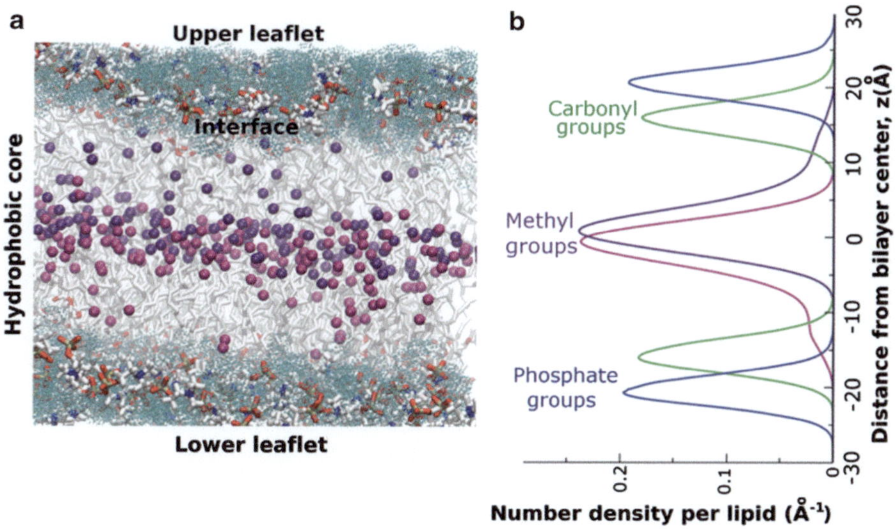

Fig. 1 (**a**) Snapshot from all-atom molecular dynamics simulations of a DOPC lipid bilayer. Lipid tails represented as *white* bonds, and the lipid heads with the color code: O—*red*, N—*blue*, P—*gold*, H—*white*. The methyl groups are represented as *purple* and *magenta* spheres corresponding, respectively, to the *upper* and *lower* leaflet of the bilayer. Water is represented as solvent-points in aquamarine. (**b**) Normalized number density for phosphate, methyl and carbonyl groups of the DOPC lipid molecules within the upper and lower leaflet. The system was prepared at 86 % hydration (7.8 waters/lipid)

such strategy, the arrangement of lipid tails in DOPC bilayers was studied [10]. Because of the difference in the scattering length of hydrogen and deuterium [11–13], the isotopic labeling on specific functional groups provides a contrast in the scattering density profiles. Synthesis of deuterium labeled DOPC lipid molecules is feasible from oleic acid specifically deuterated at the terminal methyl groups [10, 12]. In the computational set up, the neutron scattering density profile is calculated along the MD trajectory using the scattering lengths of the C, N, P, O, and H atoms [14], and then it is re-calculated using the deuterium scattering length for hydrogen atoms at the terminal methyl groups of the DOPC lipid molecules. By difference of the scattering density profiles, the location of the methyl groups along the hydrophobic core is disclosed (Fig. 2). In DOPC bilayers, broad wings in the distribution of methyl groups were detected in experiments [10], and this behavior was also observed in MD simulations using the all-atom CHARMM36 [15] or the united-atom OPLS[1] [16] force fields (Fig. 2).

Since the presence of cholesterol is linked to the formation of lipid rafts, neutron scattering experiments were designed to evaluate the structure of bilayers containing DOPC and cholesterol, at 2:1 ratio, and 7.9 water/lipid; indeed, cholesterol pro-

[1] Analysis performed on trajectories generated by Professor Martin B. Ulmschneider.

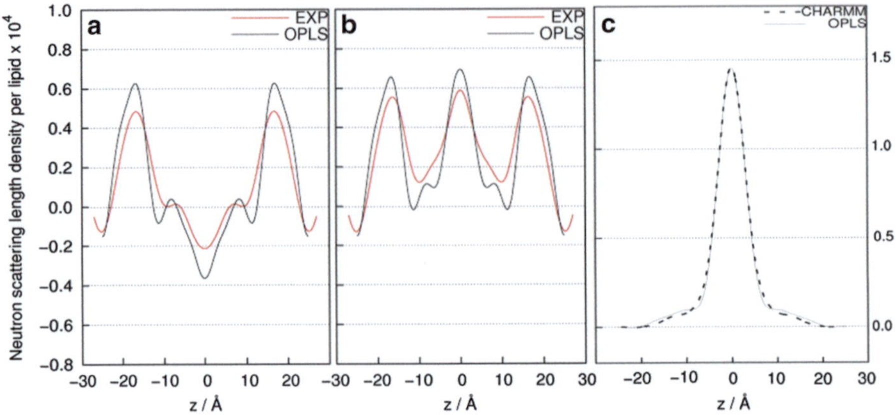

Fig. 2 Comparison of the experimental neutron scattering profile against the calculated profile from simulations using an OPLS united-atom force field for DOPC [18]. (**a**) Profile for the DOPC sample at 7.8 water/lipid. (**b**) Profiles for the deuterated DOPC sample. Scattering of terminal methyl groups are contrasted due to the larger scattering length of deuterium. (**c**) Difference of the profiles shown in panels **a** and **b**. Terminal methyl groups are located from −20 to 20 Å along the bilayer normal, and its distribution displays a maximum at z=0 Å. Experimental data from reference [10]

moted an enhancement in lipid tail ordering, and the methyl groups displayed a narrow distribution around the bilayer center [10].

By combining neutron scattering experiments and MD simulation methods, along with deuterium labeling, relevant details of the structure and function in biomembranes are being disclosed [17].

Dynamics in Lipid Bilayers

Neutron scattering experiments take advantage of the largest incoherent scattering length of deuterium (^{2}H) for studying biomembranes [19]. As hydrogen atoms are bonded to different functional groups of biomolecules, they probe the dynamics of specific domains in biomembranes [20].

The intensity detected from incoherent scattering is proportional to the spatial Fourier transform of the self-correlation function $g(r,t)$, i.e., the conditional probability of finding a particle at position r and time t, given the same particle was at position r=0 at time t=0 [20]. Since pair correlation functions can be calculated directly from MD trajectories, the scattered intensity of experimental and simulation results are comparable [9, 10].

Length and time scales accessible by neutron scattering experiments depend on the equipment's resolution: at 0.9 μeV the accessible time is ~1 ns; at 8 μeV, t~100 ps; and at 50 μeV, t~20 ps [20]. Hence, self-correlations can be determined as function of time. In a simulation, the mean square displacements MSD(t) is calculated as:

$$\mathrm{MSD}(t) = \left[\vec{r}(t) - \vec{r}(0)\right]^2 \quad (1)$$

The time dependence of the MSD shows transitions corresponding to different diffusion regimes. At $t < 10$ fs the MSD corresponds to ballistic regime; within 10 fs $< t < 30$ ns, to a cage regime; and at $t > 30$ ns, to a Fickian diffusion regime. In the ballistic regime, self-diffusion takes place within a circular cage, at the femtosecond scale; in the cage regime, the center of the circular neighborhood diffuses at the nanoseconds scale. The dynamics of LB is, therefore, described by calculating the time dependence of MSD [21].

The time dependence of MSD ($\sim t^a$) in LB display a crossover between the ballistic ($\sim t^2$) and the Fickian diffusive ($\sim t$) regime, with a typically between 0.5 and 0.7 [21, 22]. A characterization of the lipid's dynamics is important because MP's conformational changes are indeed modulated by the composition and dynamics of the LB environment [23–25]. Combining computational approaches and experimental observations is a promising strategy to decipher lipid-protein dynamical couplings, in highly complex systems such as biomembranes [20].

Dynamics of Membrane Proteins and Lipid Bilayers

Understanding the MP dynamics is a challenging task because of the interplay of fast (ps-ns) and slow (μs-ms) protein conformational changes [26] coupled to its bilayer environment [24]. In solution, globular proteins are tightly coupled to the water dynamics. The notion of water molecules controlling the fast protein conformational changes is associated to the dynamical transition, observed at ~260 K, where an abrupt increase of the protein dynamics is observed [27, 28]. At the molecular level, the protein dynamical transition is caused by the onset of water translational diffusion, which allows the protein to relax by reconfiguring the water-protein hydrogen bond network [29]. Hence, protein hydration, that extends up to 2–4 water shells [30], plays a significant role in protein dynamics. In MP, protein dynamics is even more complicated since the LB is coupled to protein function, structure, and dynamics [31].

In order to capture the dynamics of MP in bilayers, computational approaches such as MD simulations, set up in consonance with neutron scattering observables, provide important details of protein-lipid-water dynamics. For instance, MD studies on bacteriorhodopsin, a seven-helix light-driven proton pump, set up to closely correspond to the experimental samples at two hydration levels in its native purple membrane [32], reproduced the dependence of the H atom fluctuations [33]. The MSD of water molecules as function of time displayed the ballistic time dependence at short times ($\mathrm{MSD} \sim t^2$) for all the hydration levels, then a sub-diffusive regime ($\mathrm{MSD} \sim t^a$, $a < 1$), and after few ps simple diffusion ($\mathrm{MSD} \sim t$) [33]. In order to determine whether the water dynamics were coupled to the bacteriorhodopsin dynamical transition, calculations of the MSD from 150 to 300 K were performed. All systems showed an increase in the rate of change in MSD around 200 K, similar to behavior of the dynamics of protein and lipids, which was probed by the fluctuations of non-exchangeable hydrogen atoms [33]. Therefore, the protein dynamical transition in a MP coincided with signals of the onset of translational water diffusion [33]. Another interesting observation was about the dynamical transition of the purple membrane at 250 K, which was correlated with the changes in the dynamics of water molecules solvating the lipids [33]. Hence, a closer study on the role of the lipids, in particular, whether the melting of the acyl chain triggers the change of the solvation dynamics or *vice versa*, remains an open question.

Final Remarks

Understanding the dynamical couplings of MP and bilayers is work in progress. Lipid molecules resolved in the crystal structure of MP evidence the tight coupling between these components of the cell membrane. Thus, the role of lipid molecules is important for both stabilizing the protein structure in crystals and modulating the protein conformation in biomembranes. In this chapter we emphasized that molecular dynamics simulations of MP in its LB environment, designed to reproduce neutron scattering or spectroscopic data, is a compelling strategy to decipher the dynamical couplings in complex systems such as biomembranes.

References

1. Stansfeld P, Sansom MS (2011) Molecular simulation approaches to membrane proteins. Structure 19:1563–1572
2. Tan S, Tan H, Chung M (2008) Membrane proteins and membrane proteomics. Proteomics 8:3924–3932
3. Wallin E, Von Heijne G (1998) Genome-wide analysis of integral membrane proteins from eubacterial, archaean, and eukaryotic organisms. Protein Sci 7:1029–1038
4. Lingwood D, Simons K (2010) Lipid rafts as a membrane-organizing principle. Science 327:46–50
5. Jewell SA (2011) Living systems and liquid crystals. Liq Cryst 38:1699–1714
6. Veatch SL, Keller SL (2005) Seeing spots: complex phase behavior in simple membranes. Biochem Biophys Acta 1746:172–185
7. White SH, Wiener MC (1995) Determination of the structure of fluid lipid bilayer membranes. In: Disalvo EA, Simon SA (eds) Permeability and stability of lipid bilayers. CRC Press, Boca Raton, pp 1–19
8. Disalvo EA et al (2008) Structural and functional properties of hydrations and confined water in membrane interfaces. Biochim Biophys Acta 1778:2655–2670
9. Benz RW et al (2005) Experimental validation of molecular dynamics simulations of lipid bilayers: a new approach. Biophys J 88:805–817
10. Mihailescu M et al (2011) Acyl-chain methyl distributions of liquid-ordered and -disordered membranes. Biophys J 100:1455–1462

11. Blasie JK, Schoenborn BP, Zaccai G (1975) Direct methods for the analysis of lamellar neutron diffraction from oriented multilayers: a difference Patterson deconvolution approach. Brookhaven Symp Biol 27:III-58–III-67
12. Worcester DL (1988) Contrast variation and the versatility of deuterium in structural studies of biological macromolecules. J Appl Cryst 21:669–674
13. Zaccai G (2012) Straight lines of neutron scattering in biology: a review of basic controls in SANS and EINS. Eur Biophys J 41:781–787
14. Sears VF (1986) Neutron scattering lengths and cross-sections. In: Sköld K, Price DL (eds) Neutron scattering. Part A. Academic, New York, pp 521–550
15. Pastor RW, Mackerell AD (2011) Development of the CHARMM force field for lipids. J Phys Chem Lett 2:1526–1532
16. Ulmschneider JP, Ulmschneider MB (2009) United atom lipid parameters for combination with the optimized potentials for liquid simulations all-atom force field. J Chem Theory Comput 5:1803–1813
17. Mihailescu M et al (2011) Structure and dynamics of cholesterol-containing polyunsaturated lipid membranes studied by neutron diffraction and NMR. J Membr Biol 239:63–71
18. Ulmschneider MB, Sansom MSP, Di Nola A (2006) Evaluating tilt angles of membrane-associated helices: comparison of computational and NMR techniques. Biophys J 90:1650–1660
19. Peters J et al (2013) Dynamics measured by neutron scattering correlates with the organization of bioenergetics complexes in natural membranes from hyperthermophile and mesophile bacteria. Eur Phys J E 36:78
20. Zaccai G (2011) Neutron scattering perspectives for protein dynamics. J Non-Cryst Solids 357:615–621
21. Flenner E et al (2009) Subdiffusion and lateral diffusion coefficient of lipid atoms and molecules on phospholipid bilayers. Phys Rev E 79(011907):1–11
22. Khoshnood A, Jalali MA (2013) Anomalous diffusion of proteins in sheared lipid membranes. Phys Rev E 88(032705):1–7
23. Soubias O, Teague WE, Gawrisch K (2006) Evidence for specificity in lipid-rhodopsin interactions. J Biol Chem 281(44):33233–33241
24. Teague WE Jr et al (2013) Elastic properties of polyunsaturated phosphatidylethanolamines influence rhodopsin function. Faraday Discuss 161:383–395
25. Mihailescu M, Gawrisch K (2006) The structure of polyunsaturated lipid bilayers important for rhodopsin function: a neutron diffraction study. Biophys J 90(1):L04–L06
26. Henzler-Wildman KA et al (2007) A hierarchy of timescales in protein dynamics is linked to enzyme catalysis. Nat Lett 450:913–916
27. Lee AL, Wand J (2001) Microscopic origins of entropy, heat capacity and the glass transition in proteins. Nature 2001:501–504
28. Benedetto A (2013) Protein dynamics by neutron scattering. Biophys Chem 182:16–22
29. Tarek M, Tobias DJ (2002) Role of protein-water hydrogen bond dynamics in the protein dynamical transition. Phys Rev Lett 88:138101
30. Heyden M, Tobias DJ (2013) Spatial dependence of protein-water collective hydrogen-bond dynamics. Phys Rev Lett 111:218101
31. Phillips R et al (2009) Emerging roles for lipids in shaping membrane-protein function. Nature 459:379–385
32. Fitter J et al (1999) Internal molecular motions of bacteriorhodopsin: hydration-induced flexibility studied by quasielastic incoherent neutron scattering using oriented purple membranes. Proc Natl Acad Sci U S A 93:7600–7605
33. Tobias DJ, Sengupta N, Tarek M (2009) Hydration dynamics of purple membranes. Faraday Discuss 141:99–116

What Makes a Transcriptional Master Regulator? A Systems Biology Approach

Enrique Hernández-Lemus, Karol Baca-López, and Hugo Tovar

Abstract Gene regulatory networks present a wide variety of dynamical responses to intrinsic and extrinsic perturbations. An outstanding case of such coordinated responses is that of transcriptional amplification cascades, in which activation of a few key-responsive transcription factors (termed master regulators) leads to a large series of transcriptional activation events. Recent studies have pointed to the protein called *myocyte enhancing factor 2C (MEF2C)* as being one of such master regulators involved in the pathogenesis of primary breast cancer. A systems biology analysis of the transcriptional regulation activity of MEF2C and its target genes, has revealed that this molecule induces collective responses leading to system-level gene expression deregulation and carcinogenesis. We found extensive evidence for this. Being this the case, one may wonder what set of physicochemical, structural and thermodynamic constrains need to be satisfied if a protein is to become a transcription factor, and moreover a master regulator? Some hints will be discussed.

Keywords Cancer • Systems biology • Transcriptional regulation • Gene regulatory networks • Master regulators • Transcription factor • DNA binding motifs • Non-equilibrium thermodynamics • Myocyte enhancing factor 2C • Genomics of gene expression • DNA-protein interactions • Transcription factor binding site analysis

---

E. Hernández-Lemus (✉)
Center for Complexity Sciences, National Autonomous University of México, Ciudad Universitaria, 04510 México, México

National Institute of Genomic Medicine, Periferico Sur 4809, Arenal Tepepan, 14610 México, México
e-mail: ehernandez@inmegen.gob.mx

K. Baca-López
Tianguistenco Higher Studies Institute of Technology, Carretera Tenango - La Marquesa Km 22 Santiago Tilapa, Santiago Tianguistenco, México

H. Tovar
National Institute of Genomic Medicine, Periferico Sur 4809, Arenal Tepepan, 14610 México, México

L. Olivares-Quiroz et al. (eds.), *Physical Biology of Proteins and Peptides*,
DOI 10.1007/978-3-319-21687-4_10

Transcriptional Master Regulators

Gene regulatory networks present a wide variety of dynamical responses to intrinsic and extrinsic perturbations. An outstanding case of such coordinated responses is that of transcriptional amplification cascades, in which activation of a few key-responsive transcription factors (termed master regulators) leads to a large series of transcriptional activation events. Recent studies have pointed to the protein called *myocyte enhancing factor 2C (MEF2C)* as being one of such master regulators involved in the pathogenesis of primary breast cancer.

A systems biology analysis of the transcriptional regulation activity of MEF2C and its target genes, has revealed that this molecule induces collective responses leading to system-level gene expression deregulation and carcinogenesis. We found extensive evidence for this. Being this the case, one may wonder what set of physicochemical, structural and thermodynamic constrains need to be satisfied if a protein is to become a transcription factor, and moreover a master regulator? Some hints will be discussed.

Master Regulators in Gene Regulatory Networks

Phenotypic conditions in living cells are largely determined by the interplay of a multitude of molecules; in particular, genes and their protein products. The coordinated behavior of such systems is often represented as a gene regulatory network (GRN). In a GRN, regulatory processes between genes, transcription factors and other molecular components are represented by nodes and links. One common way of inferring these gene regulatory networks is by probabilistic analysis of whole genome gene expression data [9, 17].

Specific, context-dependent analysis of regulatory activity of particular cellular phenotypes (say tumor cells) may also be performed with the aid of transcriptional interaction networks. Commonly such GRNs present a complex topology, often compliant with a scale-free hierarchical structure, in which a relatively small number of key players dominate the function and dynamics of the network. Some of these key players in GRNs are transcription factors often known as *master regulators* (MRs). MRs are deemed responsible for the control of the whole regulatory program for cells under the associated phenotype [1, 5]. Master regulators may act on rather general cellular processes [12], but also on specific cellular phenotypes [1, 10, 18].

Gene regulatory network analyses have shown that different levels of gene regulation are not only related but strongly coupled. Cooperativity in transcriptional events is present, for instance, in the processes called transcriptional bursts when transcriptional cascades each due to a single promoter—or transcription factor binding event—are observed (Fig. 1).

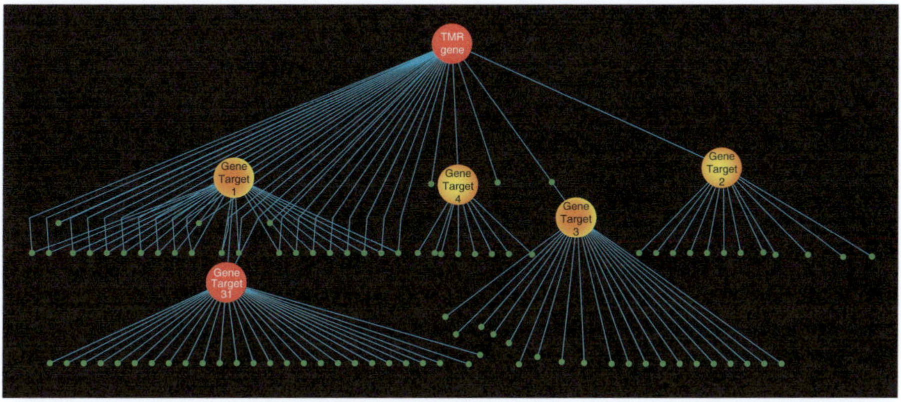

Fig. 1 A cartoon model for a TMR (big *red* node up) regulating—either directly or by means of other TFs—the transcription of more than one hundred genes. Node size is given by *out-degree*, i.e. bigger nodes have more transcriptional targets. Node *color* reflects a measure called the *closeness centrality* index, i.e. *red* nodes are closely connected with more nodes in the network whereas *green* nodes are barely connected

Some Actual Examples of Master Regulators

For instance, it is known that mTOR is active in concerting signals regulating control growth, metabolism, and longevity. Malfunction of mTOR complexes has been associated with developmental abnormalities, autoimmune diseases and cancer [12]. The main role of mTOR seems to be the regulation of protein synthesis. Detailed mechanisms remain unknown, but ribosome profiling seems to point to translational regulation and transcriptional activation activity. Due to the multiplicity of mTOR signaling interactions, this molecule acts as a master regulator on a variety of phenotype conditions.

Specific master regulatory activity may be exemplified by cases such as that of VASH1 that has been identified as a master-regulator of endothelial cell apoptosis [1]; PAX5 is known to be a master regulator of B-cell development also involved in neoplastic processes in leukemogenesis [18] and the yeast protein Gcn4Pp (that contains a conserved domain cd12193 present in human JUN proto-oncogene) that is triggered by starvation and stress signals [10] and is an MR in the phenotypic response to such stimuli. Due to the complex mechanisms behind transcriptional regulation in eukaryotes, identification of MRs is mostly based on the (inferred or empirical) relationships between them and their downstream RNA targets in the GRN.

Transcriptional master regulators in eukaryotes commonly operate in a *phenotype-dependent fashion*, unlike global transcriptional regulators in prokaryotes.

The MEF2 Family of Transcription Factors as Novel Cancer-Associated Transcriptional Master Regulators

As may be evident from the few examples just mentioned, MRs may have a big impact on cancer-related phenotypes. This is so since under genome instability conditions, the uncontrolled synthesis of these molecules may give rise to large amplification of transcriptional cascades. In [4] the role that some molecules of the family of MEF2 transcription factors (in particular, MEF2C) may have in processes involving metabolic deregulation and MR activity at the onset of primary breast cancer was studied. The approach followed there involved the inference of GRNs centered in a number of molecules considered to be candidate MRs associated with the breast cancer phenotypes at early stages (i.e. primary tumors). As is seen there, MEF2C showed to be a quite promising molecule due to the large number of (probabilistically inferred) targets that it possesses, but also due to the main biological processes spanned by its targets.

A connection exists between energy deregulation at the level of cell metabolism and global genomic instability as triggered by transcriptional bursts of activity. The action of MRs may thus be able to trigger transcriptional cascades in which a big number and variety of mRNA transcripts are synthesized in a series of interrelated events. This may happen due to the fact that TFs (in particular MRs) often possess quite low *activation energies*, i.e. the minimum amount of energy needed to carry on their cellular process of synthesis, given the necessary nucleotides (for transcription) and aminoacids (for translation). Activation energies are a measure of how likely is a chemical reaction—or a series of chemical reactions—to occur. The lower the activation energy, the higher the probability of the TF synthesis to happen. MR cascading phenomena is important for certain stances in cell development, differentiation and proliferation, but their abnormal function could lead the cells into neoplastic states. Baca-López et al. [4] find out that non- equilibrium free energies calculated at the level given by extended irreversible thermodynamics [8], provide a realistic description of transcription factor activation in the case of MRs.

> Imagine you have a cell with deregulated metabolism (e.g. **large local free energy fluctuations**), a **TF** gene with **low activation energy** located close to energy transduction pathways, with a **high hierarchy** on the transcriptional regulation network and a relatively **large relaxation time for mRNA decay**.

The Coordinated Response of MEF2C Targets Induces Global Phenotypic Patterns: Gene Expression Signatures Characteristic of Breast Cancer Phenotypes

As was already mentioned, an interesting feature of the behavior of most MR in eukaryotes is the fact that master regulatory activity seems to be phenotype-specific, unlike global regulators in prokaryotes that often dominate transcriptional network

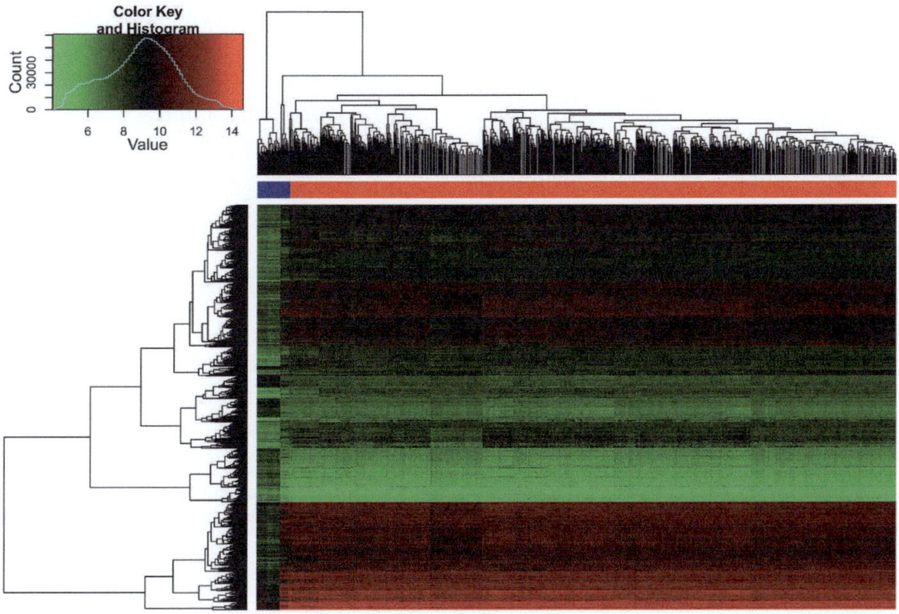

Fig. 2 The plot presents a Heatmap for gene expression hierarchical clustering among MEF2C and its targets up to third-order in primary breast carcinomas vs healthy tissue. The figure shows that genes in the MEF2C-dependent transcriptional subnetwork present differences in gene expression patterns between cancer and normal tissue that may account for MEF2C as a master regulator for the transition between normal tissue and neoplasms

dynamics for every phenotype. MEF2C seem to be an MR in cancer-related phenotypes. This is in accordance with the fact that MEF2C primary functions are related to proliferative cell behavior. Indeed, a number of molecules in the so-called *proliferative cluster* [6] are direct (sometimes indirect) transcriptional targets of MEF2C. Let us see how MEF2C regulated genes are sufficient to *induce the molecular phenotype* in a case/control comparison, i.e. to mark clearly distinguishable expression patterns between tumors and healthy cells.

In Fig. 2, a heatmap of the bi-clustering of genes and samples of differentially expressed genes that are transcriptional targets of MEF2C. We can see that genes transcriptionally regulated by MEF2C or its targets, are able to *induce the phenotype* that is, changes in the expression of such genes correlate with phenotypic changes. Such behavior is reminiscent of that of molecular profiling of gene expression samples [28]. Figure 2 can be read (and interpreted) as follows: Every row is a gene for the collection of MEF2C targets and every column is a sample for a set of 880 whole genome gene expression experiments. The color range refers to the *z-score* statistic for the distribution (in the inset at the left superior corner) of the so-called *log-fold change* which is the base 2 logarithm of the ratio between the levels of gene expression of tumors/controls for a single gene. Green values in a row/column cell indicate that the given gene is *underexpressed* in tumors as compared to controls and red is the other way around.

The heatmap shows that the gene expression patterns reflect characteristic ways in which genes are transcriptionally regulated by the action of a MR and how, these patterns are consistent enough to distinguish between tumors (red bar over the heatmap) and controls (blue bar over the heatmap). Genes and samples were sorted following an unsupervised hierarchical clustering as shown by the dendrograms in the upper and left side of the heatmap plot.

The MEF2 Family and Their Repertoire of DNA-Binding Domains

MEF2C is a transcription factor gene located (in humans) in 5q14.3 on the minus strand. This gene is 200,723 bp long and encodes a 473-aminoacid protein weighting 51.221 kDa. MEF2C is a member of the Mef2 family that by means of controlling gene expression (MEF2 molecules are commonly acting as activator transcription factors) is able to regulate cellular differentiation and development [24]. MEF2 members are highly versatile regulators since they contain both **MEf2** and **MADS-box** DNA binding domains. The MADS-box serves as the minimal DNA-binding domain; however, an adjacent 29-amino acid extension called the Mef2 domain is required for high affinity DNA-binding and dimerization, hence conferring a combinatorial DNA binding repertoire through a number of transcription factor recognition marks (see Figs. 3, 4 and 5) [19].

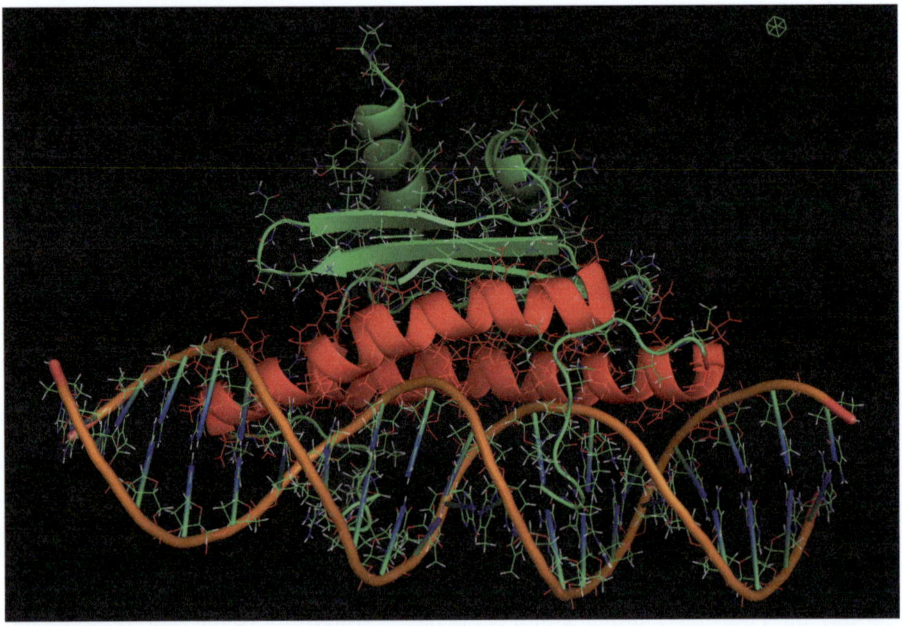

Fig. 3 NMR-resolved molecular **PyMol** visualization of the MEF2A MADS DNA-binding core attached to a 20mer consensus sequence CTA(A/T)(4)TAG [**PDB Key: 1C7U**]. For pedagogical purposes MADS DNA-binding domain is colored in *red*

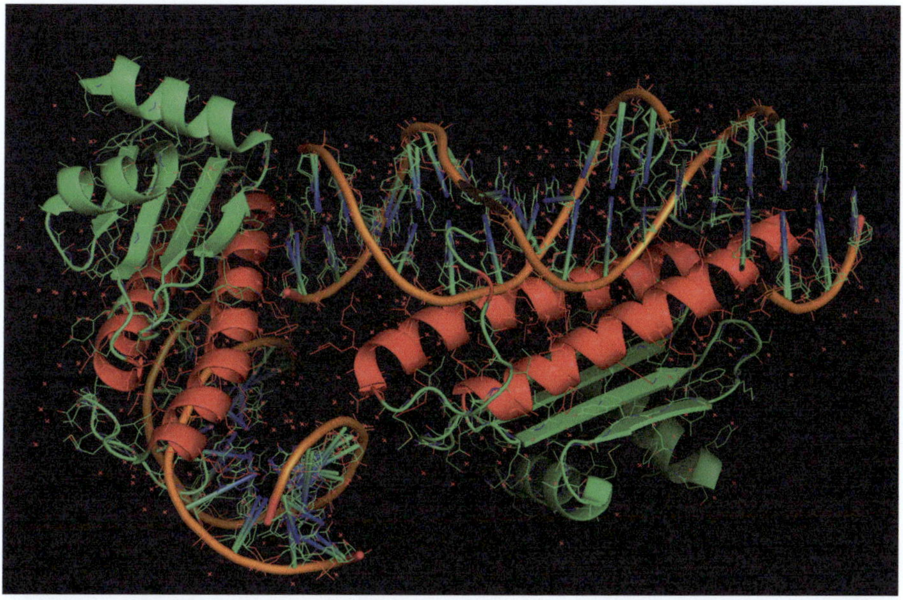

Fig. 4 Crystal structure molecular **PyMol** visualization of 2 copies of the protein MEF2A with their MADS DNA-binding cores attached to a consensus sequence [**PDB Key: 1EGW**]. Experimental resolution was 1.5 Å. For pedagogical purposes MADS DNA-binding domains are colored in red. We can notice that the MADS domains in this transcription factor protein are able to bind DNA in different spacial configurations

Furthermore, it is known that the MEF2C protein interacts with MAPK7 (involved in proliferation and differentiation signaling) [34], EP300 (a transcription factor that regulates cell growth and cellular division) [25], TEAD1 (an enhancer TF that co-regulates transcription with MEF2C) [7], as well as with a number of histone deacetylases, most notably HDAC4, HDAC7, and HDAC9 [30, 31]. These protein-protein interactions, mostly with other transcription factors, enhancers or epigenomic regulators, together with their inherent binding-site transcriptional activity make MEF2C a functional and adaptable MR.

A **DNA-binding domain** (DBD) is a conserved part of a protein sequence that has an independently-folded and preserved spatial structure, and that in addition includes at least one compact (supersecondary) structure or *motif* that is able to *recognize* and bind DNA. Recognition and binding by DBD is based on non-covalent electrostatic interactions such as hydrogen and halogen bonds, hydrophobic and van der Waals forces, and $\pi - \pi$ interactions leading to molecular complementarity.

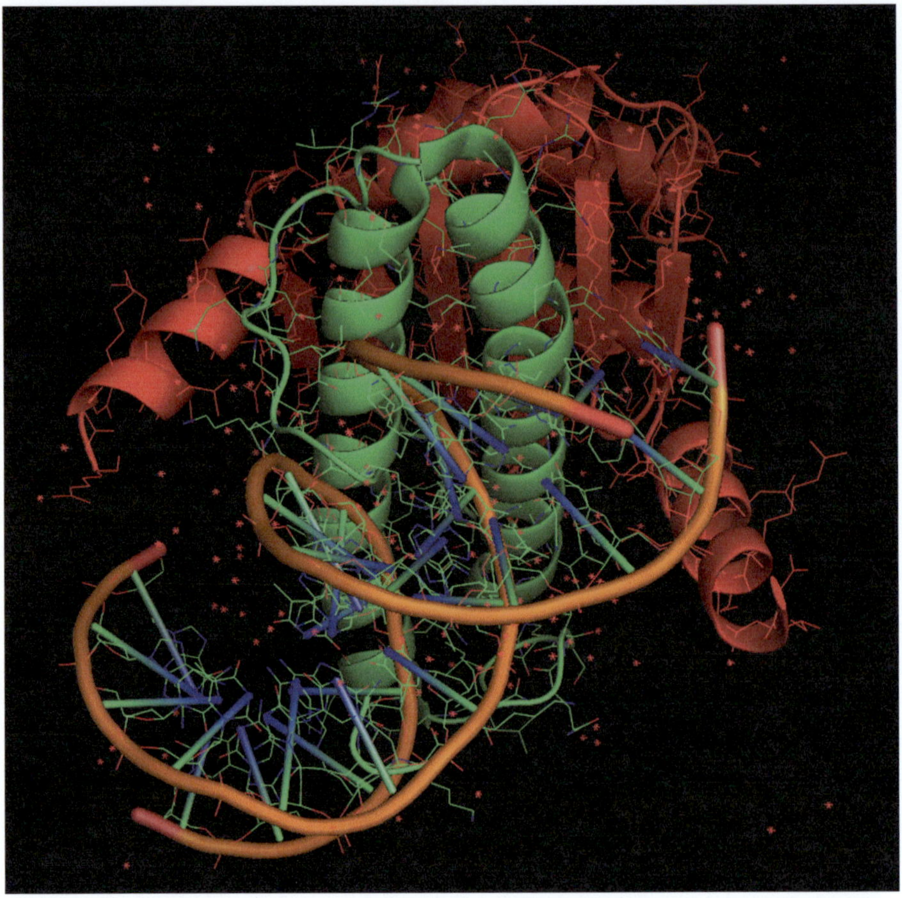

Fig. 5 Crystal structure molecular **PyMol** visualization of the protein MEF2B with their MEf2 DNA-binding core attached to a consensus sequence [**PDB Key: 1N6J**]. For pedagogical purposes MEf2 DNA-binding domain is colored in *red*

Master transcriptional regulators have often conserved usage across species. In the particular case of MEF2C this is not surprising due to the characteristic sequence structure that it presents. In fact, it is very likely that MEF2C owes its master regulatory effects to its MADS-box structure. It has been shown that the MADS-box gene transcription factor motif that is extremely conserved across plant and animal kingdoms [2]. As an extreme example, in [4] we analyzed the comparative genomics of this motif in *Homo sapiens* (H.s.) and *Arabidopsis thaliana* (A.t.). In that reference, a representative short aminoacid sequence (60 aa) from MEF2C human protein (MEF2C_Hs) was taken from [2] and used for searching conserved domain sequences within A.t. MADs family members, by using the NCBI Blast protein tool. Protein conserved domains from putative

MADS-box family transcription and Mef2 myocyte enhancer factor 2 families sharing domain-architecture were found, 50 (A.t.) and 101 (H.s.) respectively ($E_{value} < 1 \times 10^{-07}$).

The MADS-box structure is a highly conserved DNA-binding motif present in most of the eukaryotes. The MADS-box domain is able to bind to the **CArG-box** a very common motif present in a large number of promoter regions of the eukaryotic genome. The typical length of the MADS-box is around 168–180 bps (or 56–60 amino acids). In animals and fungi MEF2-like proteins have a second DBD called MEf2. By means of an interaction with the MADS-box the family of MEF2 transcription factors are able to dimerize (forming either homodimers or heterodimers) and reach nuclear localization.

Known Roles of MEF2C in Humans

Abnormal function of the MEF2C molecule has previously been associated with disease, mostly related to neural and muscular disorders [22]. However, due to the fact that MEF2C is an active player in the cell proliferation and differentiation processes, its importance in the pathogenesis of cancer should not be overlooked; to date, most of the evidence has been found in leukemia [11]. Homminga et al. [11] mentions MEF2C as a potential oncogene since it is an activator of oncogenic pathways in T cell acute lymphoblastic leukemia (T-ALL). In that study, by applying prediction analysis for microarray (PAM) expression data, it was possible to characterize MEF2C as being consistently over expressed in the so-called *immature phenotype* cluster of samples which proliferate while lacking known oncogenic rearrangements. By combining expression analysis with molecular cytogenetics, it was possible to identify chromosomal abnormalities in T-ALL converging to MEF2C activation.

For instance, translocation of BCL11B to SPI1 has been found to lead to high expression of MEF2C as is the case with gene fusion products of RUNX1 such as RUNX1/AML1. MEF2C transfection experiments also contribute evidence that MEF2C may function as a transcriptional regulator in immature T-ALL leading to the conclusion that activation of LMO2 and LYL1 by MEF2C could be a fundamental event for the gathering of early-committed T-cells in leukemogenesis [11]. The oncogenic capacity of MEF2C to transform cells was tested and the results show that while MEF2C alone was insufficient to drive cellular transformation in NIH 3T3 assays, it was indeed able to synergize with RAS or MYC to start cellular transformation. With all these tests analyzed, early conclusions suggest that MEF2C may be a central oncogene for immature T-ALL cases [11].

MEF2C is also involved in other proliferation and cell cycle processes that may be potentially oncogenic. For instance, it is known that BMK1 (a MAP kinase activated by a number of growth factors) is required to trigger growth factor induced cell proliferation. BMK1 regulates the expression of c-JUN, a well-known proto-oncogene, that is also an important regulator of gene expression- by means of direct phosphorylation and activation of MEF2C [15]. Proliferative and cell cycle activity are not the only potentially-oncogenic features of MEF2C over regulation; it has been demonstrated that high levels of MEF2C are able to inhibit the expression of NR4A1/NUR77 that regulates apoptosis by means of BCL2 transformation [20].

MEF2C participates in the regulation of the retinoblastoma tumor suppressor protein (pRB) [21], a critical regulator of late stages of cell cycle/gene expression/differentiation processes, and it is also involved in ErbB1 downstream signaling by positive regulation of ERK5 [29] mediating growth factor responses in a process involving the Raf proto-oncogene [13] as well as the MAP kinase BMK1 [14]. It is known that liver regeneration phosphatase (PRL-3) is overexpressed in a number of tumors and also that its transcript level is high in the vascular and endothelial cells of malignant tissue. Recent studies suggest that PRL-3 functions downstream in the MEF2C regulated VEGF pathway in endothelial cells, related to tumor angiogenesis [33]. Interestingly, the anticancer drugs Doxorubicin and Resveratrol have been demonstrated to decrease the expression of MEF2C at the mRNA level in murine models [23, 26].

Transcription Factor Binding Site Analysis: Physical Basis and Biological Consequences

Transcription factor binding site analysis is the generic name of a set of extremely important methodologies in contemporary computational biology. The theoretical basis of TFBS lies in the heart of Physical Biology, encompassing powerful ideas from electrostatic and electrodynamical modeling, structural biology, thermodynamics, chemical kinetics and so on. The reliability and predictive powers of TFBS may be greatly enhanced by taking into consideration not only local molecular interaction patterns, but also the fact that multiple transcription factors may be able to bind nearby sites, the spatial clustering of TFs in so-called *cis*-regulatory modules and motif conservation across orthologues [3].

Physicochemical interactions behind the binding of transcription factors and DNA is—as we are already aware—of foremost importance for transcriptional regulation (TR). TR is, however a process that involves more than just static interactions: environmental and tissue specific constraints are also playing important roles. Electrostatic interactions involve not only DNA genomic regions and DBDs but also the presence of additional transcription factor proteins, chromatin remodeling and histone modifications.

The systematic identification of transcription factor binding sites is a very important area of research. Predictive models reflecting the relative DNA binding site and nucleotide preference in a positional context are quantitatively described by means of the so-called *Position Weight Matrix* (PWM) [27]. PWMs are based on empirical evidence coming from sequence specific DNA-binding experiments, electrostatic calculations both classic and semi-classic, conserved sequence alignment and probabilistic models for cooperativity between different TFs [32].

Position Weight Matrices (PWMs) reflect the relative importance of several physical constraints in the physicochemical microenvironment that may allow—or not—for a transcription factor to bind to DNA. These factors include (among others):

- Electrostatic interactions strong enough to support TF-DNA binding. A more stable electrostatic pairing interaction is favored with a higher PWM score.
- The appropriated sequence-specific DNA conformation: i.e. a TF that binds to an *open* DNA segment (i.e. one with no steric constrains for the spatial coupling) will be favored when compared to one *closed*. An easier to bind configuration match will get a higher PWM score.
- A lower local absolute chemical potential (i.e. one in which histone modifications and the presence of other TF proteins stabilize the chemical microenvironment) will grant also higher PWM scores.

TFBS binding specificities can be further assessed and validated experimentally. High-throughput chromatin immunoprecipitation with sequencing (ChIP-seq) is likely the preferred experimental strategy to validate protein-DNA interactions. However CHIP-seq is not able to resolve TF-DNA binding at base-pair resolution level. Further improvement on this has been explored by Udalova et al. [16] used a microarray-based quantitative protein-DNA binding assay as well as Surface Plasmon Resonance to determine the relative affinities and the association/dissociation rate constants respectively in an optimized selection of DNA variants. Chemical affinity and dissociation analyses thus have the potential to reveal the real strength of intermolecular TFBS interactions.

While it is true that from purely electrostatic arguments identical genomic sequences should have the same affinity for a given TF, experimental observations have shown that certain selected sequences possess a distinctively high affinity playing the role of functional TFBS, very likely due to structural and conformational constraints such as chromatin remodeling patterns. Biophysical considerations have also lead to consider the cooperative contribution of many relatively *weak* binding sites as well as the competition of several TFs targeting the same sites by accounting for all non-overlapping binding sites in a combinatorial procedure. The `MotEvo` algorithm considers these biophysical arguments by means of a Bayesian approach. The input data for the algorithm is a set of one or more PWMs supplemented with phylogenetic and functional information as secondary priors [3].

Concluding Remarks on the Physics and Biology Behind a Master Regulator

In this chapter we have discussed a number of features that characterize a transcriptional master regulator. Such features involve phenomena happening both at the biophysical level and at the purely mechanistic biological function level also. A view based on the approach of systems biology (integrating information from a number of high-throughput sources of both biophysical and genomic data) included considerations in the MR protein itself, such as the fact that a MR is so, among other things due to electrostatic, conformational and thermodynamic reasons. Also the functional roles of MRs in the cell under normal and abnormal conditions is discussed.

- Low activation energy (that is but negatively regulated by a large number of miRNAs
- Versatile mechanisms of DNA binding by diverse Transcription Factor Binding Site patterns
- High number of regulated molecules both directly and indirectly
- Plausibility to show cascading phenomena

By using the concrete example of the MEF family of myocyte enhancing factors (in particular focusing on the MEF2C member that we have been studying in our lab for some years) we pointed out to the interplay of physicochemical approaches, computational structural biology and bioinformatics as well as whole genome transcriptomics as it can be seen in Table 1.

Transcriptional regulation programmes are of course extremely complex to deal with, particularly in the highly variable conditions found in neoplastic diseases. However, we are convinced that by carrying out systematic analyses able to combine the solid theoretical insight provided by the physical sciences with the vast amount of data present in contemporary biology and biomedical settings it is possible to further improve of own understanding of the role of transcriptional regulators, both in normal settings—e.g. in developmental processes or aging—and in pathological conditions. The road ahead is still long and unknown, we believe, however, that further studies along these lines will take us on into the right track.

Table 1 A number of **MEF2C** features implied on its transcriptional master regulator activity

MEF2C TMR parameters	Value
Activation energy	472.81 Kcal/mol
Number of regulating miRNAs	572
TFBS mechanisms	DBD of two quite general classes: MADS and MEf2
Number of potential first order regulatory interactions[a]	158
Number of potential second order regulatory interactions[a]	1308
Number of potential third order regulatory interactions[a]	4550
Target genes with TF activity	CREB3, KLF4, CEBPB, HMGA2, ZIC2, FOXN3, etc.

[a] As inferred by the `MotEvo` algorithm in the `FANTOM4` database

Acknowledgements This work was supported by CONACyT (Grant 179431/2012) as well as federal funding from the National Institute of Genomic Medicine (Mexico). Additional support from the National Laboratory of Complexity Sciences is also acknowledged.

References

1. Affara M, Sanders D, Araki H, Tamada Y, Dunmore BJ, Humphreys S, Imoto S, Savoie C, Miyano S, Kuhara S, Jeffries D, Print C, Charnock-Jones DS (2013) Vasohibin-1 is identified as a master-regulator of endothelial cell apoptosis using gene network analysis. BMC Genomics 14:23
2. Álvarez-Buylla ER, Pelaz S, Liljegren SJ, Gold SE, Burgeff C, Ditta GS, Ribas de Pouplana L, Martínez-Castilla L, Yanofsky MF (2000) An ancestral MADS-box gene duplication occurred before the divergence of plants and animals. Proc Natl Acad Sci USA 97(10): 5328–5333
3. Arnold P, Erb I, Pachkov M, MOlina N, van Nimwegen E (2012) MotEvo: integrated Bayesian probabilistic methods for inferring regulatory sites and motifs on multiple alignments of DNA sequences. Bioinformatics 28(4):487–494
4. Baca-López K, Hidalgo-Miranda A, Mayorga M, Gutiérrez-Nájera N, Hernández-Lemus E (2012) The role of master regulators in the metabolic/transcriptional coupling in breast carcinomas. PLoS ONE. doi:10.1371/journal.pone.0042678
5. Basso K, Margolin AA, Stolovitzky G, Klein U, Dalla-Favera R, Califano A (2005) Reverse engineering of regulatory networks in human B cells. Nat Genet 37(4):382–390
6. Brosh R, Rotter V (2010) Transcriptional control of the proliferation cluster by the tumor suppressor p53. Mol Biosyst 6(1):17–29
7. Fukuda K (2002) Molecular characterization of regenerated cardiomyocytes derived from adult mesenchymal stem cells. Congenit Anom 42(1):1–9
8. Hernández-Lemus E (2009) Non-equilibrium thermodynamics of gene expression and transcriptional regulation. J Non-Equilib Thermodyn 34(4):371–394
9. Hernández-Lemus E, Velázquez-Fernández D, Estrada-Gil JK, Silva-Zolezzi I, Herrera-Hernández MF, Jiménez-Sánchez G (2009) Information theoretical methods to deconvolute genetic regulatory networks applied to thyroid neoplasms. Phys A 388:5057–5069
10. Hinnebusch AG, Natarajan K (2002) Gcn4p, a master regulator of gene expression, is controlled at multiple levels by diverse signals of starvation and stress. Eukaryot Cell **1**:22–32
11. Homminga I, Pieters R, Langerak AW, de Rooi JJ, Stubbs A, Verstegen M, Vuerhard M, Buijs-Gladdines, J, Kooi C, Klous P, van Vlierberghe P, Ferrando AA, Cayuela JM, Verhaaf B, Beverloo HB, Horstmann M, de Haas V, Wiekmeijer AS, Pike-Overzet K, Staal FJ, de Laat W, Soulier J, Sigaux F, Meijerink JP (2011) Integrated transcript and genome analyses reveal NKX2-1 and MEF2C as potential oncogenes in T cell acute lymphoblastic leukemia. Cancer Cell 19(4):484–497
12. Hosking R (2012) mTOR: the master regulator. Cell 149:955–957
13. Janulis, M, Trakul N, Greene G, Schaefer EM, Lee JD, Rosner MR (2001) A novel mitogen-activated protein kinase is responsive to Raf and mediates growth factor specificity. Mol Cell Biol 21(6):2235–2247
14. Kato Y, Kravchenko VV, Tapping RI, Han J, Ulevitch RJ, Lee JD (1997) BMK1/ERK5 regulates serum-induced early gene expression through transcription factor MEF2C. EMBO J 16(23):7054–7066
15. Kato Y, Zhao M, Morikawa A, Sugiyama T, Chakravortty D, Koide N, Yoshida T, Tapping RI, Yang Y, Yokochi T, Lee JD (2000) Big mitogen-activated kinase regulates multiple members of the MEF2 protein family. J Biol Chem 275(24):18534–18540
16. Linnell J, Mott R, Field S, Kwiatkowski DP, Ragoussis J, Udalova IA (2004) Quantitative high-throughput analysis of transcription factor binding specificities. Nucleic Acids Res 32:4

17. Margolin AA, Nemenman I, Basso K, Wiggins C, Stolovitzky G, Dalla FR, Califano A (2006) ARACNe: an algorithm for the reconstruction of gene regulatory networks in a mammalian cellular context. BMC Bioinf 7(Suppl I S7):1–15
18. Medvedovic J, Ebert A, Tagoh H, Busslinger M (2011) Pax5: a master regulator of B cell development and leukemogenesis. Adv Immunol 111:179–206
19. Molkentin JD, Olson EN (1996) Combinatorial control of muscle development by basic helix-loop-helix and MADS-box transcription factors. Proc Natl Acad Sci USA 93(18):9366–9373
20. Nagel S, Meyer C, Quentmeier H, Kaufmann M, Drexler HG, MacLeod RA (2008) MEF2C is activated by multiple mechanisms in a subset of T-acute lymphoblastic leukemia cell lines. Leukemia 22(3):600–607
21. Novitch BG, Spicer DB, Kim PS, Cheung WL, Lassar AB (1999) pRb is required for MEF2-dependent gene expression as well as cell-cycle arrest during skeletal muscle differentiation. Curr Biol 9(9):449–459
22. Nowakowska BA, Obersztyn E, Szyman'ska K, Bekiesin'ska-Figatowska M, Xia Z, Ricks CB, Bocian E, Stockton DW, Szczauba K, Nawara M, Patel A, Scott DA, Cheung SW, Bohan TP, Stankiewicz P (2010) Severe mental retardation, seizures, and hypotonia due to deletions of MEF2C. Am J Med Genet B Neuropsychiatr Genet 153B(5):1042–1051
23. Poizat C, Sartorelli V, Chung G, Kloner RA, Kedes L (2000) Proteasome-mediated degradation of the coactivator p300 impairs cardiac transcription. Mol Cell Biol 20(23):8643–8654
24. Potthoff MJ, Olson EN (2007) MEF2: a central regulator of diverse developmental programs. Development 134(23):4131–4140
25. Sartorelli V, Huang J, Hamamori Y, Kedes L (1997) Molecular mechanisms of myogenic coactivation by p300: direct interaction with the activation domain of MyoD and with the MADS box of MEF2C. Mol Cell Biol 17(2):1010–1026
26. Smeding L, Leong-Poi H, Hu P, Shan Y, Haitsma JJ, Horvath E, Furmli S, Masoom H, Kuiper JW, Slutsky AS, Parker TG, Plotz FB, dos Santos CC (2012) Salutary effect of resveratrol on sepsis-induced myocardial depression. Crit Care Med 40(6):1896–1907
27. Stormo GD (2000) DNA binding sites: representation and discovery. Bioinformatics 16:16–23
28. Subramanian A, Tamayo P, Mootha KV, Mukherjee S, Ebert BL, Gillette MA, Paulovichg A, Pomeroyh SL, Golub TR, Landera ES, Mesirov JP (2005) Gene set enrichment analysis: a knowledge-based approach for interpreting genome-wide expression profiles. Proc Natl Acad Sci USA 102(43):15545–15550
29. Sun W, Wei X, Kesavan K, Garrington TP, Fan R, Mei J, Anderson SM, Gelfand EW, Johnson GL (2003) MEK kinase 2 and the adaptor protein Lad regulate extracellular signal-regulated kinase 5 activation by epidermal growth factor via Src. Mol Cell Biol 23(7):2298–2308
30. Wang AH, Bertos NR, Vezmar M, Pelletier N, Crosato M, Heng HH, Th'ng J, Han J, Yang XJ (1999) HDAC4, a Human histone deacetylase related to yeast HDA1, is a transcriptional corepressor. Mol Cell Biol 19(11):7816–7827
31. Wang AH, Yang XJ (2001) Histone deacetylase 4 possesses intrinsic nuclear import and export signals. Mol Cell Biol 21(17):5992–6005
32. Whitfield TW, Wang J, Collins PJ, Partridge EC, Aldred SF, Trinklein ND, Myers RM, Weng Z (2012) Functional analysis of transcription factor binding sites in human promoters. Genome Biol 13:R50
33. Xu J, Cao S, Wang L, Xu R, Chen G, Xu Q (2011) VEGF promotes the transcription of the human PRL-3 gene in HUVEC through transcription factor MEF2C. PLoS ONE 6:11
34. Yang CC, Ornatsky OI, McDermott JC, Cruz TF, Prody CA (1998) Interaction of myocyte enhancer factor 2 (MEF2) with a mitogen-activated protein kinase, ERK5/BMK1. Nucleic Acids Res 26(20):4771–4777

Index

L. Olivares-Quiroz et al. (eds.), *Physical Biology of Proteins and Peptides*,
DOI 10.1007/978-3-319-21687-4

MIX
Papier aus verantwortungsvollen Quellen
Paper from responsible sources
FSC® C105338

If you have any concerns about our products,
you can contact us on
ProductSafety@springernature.com

In case Publisher is established outside the EU,
the EU authorized representative is:
Springer Nature Customer Service Center GmbH
Europaplatz 3, 69115 Heidelberg, Germany

Printed by Libri Plureos GmbH
in Hamburg, Germany